CREATING BEAUTIFUL FLOORS

By the Editors of Sunset Books

Sunset Books

President and Publisher: Susan J. Maruyama
Director, Sales & Marketing: Richard A. Smeby
Marketing & Creative Services Manager: Guy C. Joy
Production Director: Lory Day
Editorial Director: Bob Doyle

Sunset Publishing Corporation

Chairman: Jim Nelson
President/Chief Executive Officer: Stephen J. Seabolt
Chief Financial Officer: James E. Mitchell
Publisher: Anthony P. Glaves
Circulation Director: Robert I. Gursha
Director of Finance: Larry Diamond
Vice President, Manufacturing: Lorinda B. Reichert
Editor, Sunset Magazine: William R. Marken

Creating Beautiful Floors was produced in conjunction with

St. Remy Press

President/Chief Executive Officer: Fernand Lecoq
President/Chief Operating Officer: Pierre Léveillé
Vice President, Finance: Natalie Watanabe
Managing Editor: Carolyn Jackson
Managing Art Director: Diane Denoncourt
Production Manager: Michelle Turbide

Editorial staff for *Creating Beautiful Floors*

Senior Editor: Heather Mills
Editor: Alfred LeMaitre
Senior Art Director: Francine Lemieux
Art Director: Solange Laberge

Book Consultant

Don Vandervort

Special Contributors

Eric Beaulieu, Michel Blais, Normand Boudreault, Caroline Bowden, Marc Cassini, Robert Chartier, François Daxhelet, Hélène Dion, Jean-Guy Doiron, Lorraine Doré, Dominique Gagné, Michel Giguère, Christine M. Jacobs, François Longpré, Geneviève Monette, Jennifer Ormston, Jacques Perrault, Adam van Sertima, Rebecca Smollett, Judy Yelon

COVER: Interior design by Diane Davis, Foster City, California. Cover Design by Kevin Freeland. Photography by Philip Harvey. Photo Direction by JoAnn Masaoka Van Atta.

First printing January 1996
Copyright © 1996
Published by Sunset Publishing Corporation,
  Menlo Park, CA 94025.

ISBN 0-376-01147-5
Library of Congress Catalog Card Number: 95-070326
Printed in the United States

For more information on *Creating Beautiful Floors* or any other Sunset Book, call 1-800-634-3095. For special sales, bulk orders, and premium sales information, call Sunset Custom Publishing Services at (415) 324-5577.

Acknowledgments

Thanks to the following:
American Olean Tile Company, Lansdale, PA
American Wood Preservers Association, Woodstock, MD
APA-The Engineered Wood Association, Tacoma, WA
Armstrong World Industries Canada Ltd., Montreal, Que.
Azrock Industries Inc., San Antonio, TX
Brick Institute of America, Reston, VA
Burke Flooring Products, San Jose, CA
Carpet and Rug Institute, Dalton, GA
Congoleum Corporation, Mercerville, NJ
Crossville Ceramics, Bedford, TX
Richard Day, Palomar Mountain, CA
Domco Industries Ltd., Farnham, Que.
DuPont Canada Inc., Mississauga, Ont.
Floor Covering Installation Contractors Association,
  Dalton, GA
Harris-Tarkett Inc., Johnson City, TN
Joy Carpets Inc., Fort Oglethorpe, GA
Kentucky WoodFloors, Louisville, KY
Laticrete International Inc., Aurora, Ont. and Bethany, CT
Mannington Mills Inc., Brampton, Ont.
Masland Carpets Inc., Mobile, AL
Giles Miller-Mead, Brome, Que.
Mitchell and Company Inc., Dalton, GA
National Floor Products Company Inc., Florence, AL
National Oak Flooring Manufacturers Association,
  Memphis, TN
National Wood Flooring Institute, Manchester, MO
Resilient Floor Covering Institute, Rockville, MD
Stern and Associates, Cranford, NJ
Tile Council of America, Clemson, SC
Weightman Public Relations, Philadelphia, PA
World Floorcovering Association, Anaheim, CA
W.W. Henry Co., Huntington Park, CA

Picture Credits

p. 6   Rob Super
p. 7   *(both)* Steve W. Marley
p. 8   *(upper)* courtesy Kentucky WoodFloors
p. 8   *(lower)* Steve W. Marley
p. 9   courtesy Armstrong World Industries Canada Ltd.
p. 10 *(upper)* Steve W. Marley
p. 10 *(lower)* courtesy Congoleum Corporation
p. 11 *(upper)* courtesy Mannington Mills Inc.
p. 11 *(lower)* courtesy Congoleum Corporation
p. 12 Steve W. Marley
p. 13 *(upper)* Rob Super
p. 13 *(lower)* Steve W. Marley
p. 14 *(upper)* Rob Super
p. 14 *(lower)* courtesy American Olean Tile Co.
p. 15 Jack McDowell
p. 16 *(upper)* Jack McDowell
p. 16 *(lower)* Tom Wyatt
p. 17 *(left)* courtesy American Olean Tile Co.
p. 17 *(right)* Steve W. Marley
p. 18 courtesy DuPont Canada Inc.
p. 19 *(both)* courtesy DuPont Canada Inc.
p. 20 *(upper)* courtesy Joy Carpets Inc.
p. 20 *(lower)* courtesy Masland Carpets Inc.

CONTENTS

GETTING STARTED IN FLOORING

Wood, resilient flooring, ceramic tile, masonry, and carpeting are all available today in a wide variety of styles and qualities, and are all good flooring choices. New flooring materials, effective adhesives, and durable sealers and finishes make flooring easy to install and maintain.

To become familiar with the different types of flooring available and to choose one that both complements the interior design of your home and meets your needs, read the descriptions of different types of flooring beginning on page 5; you'll discover innovative uses of floor coverings that will spark your imagination.

Using basic flooring materials and a little imagination, you can create special flooring effects that are interesting and unusual, and may have practical merits, as well: Squares of resilient tile can be arranged to create a giant checkerboard on a game room floor; smaller ceramic tiles can produce strikingly colorful patterns in custom designs. Industrial floor coverings are finding acceptance in homes, where their toughness is considered an advantage in high-traffic areas.

Before you begin your project, you'll have to thoroughly inspect your existing floor and make sure there are no underlying structural problems *(page 21)*. You should also fix any squeaky floors *(page 22)* or stairs *(page 25)* before you install any flooring material or carpeting.

Use a straightedge to find dips in the floor (page 21). *It's an important first step in preparing to install flooring material, and may also help you discover any structural problems.*

FLOORING GUIDELINES

Innovations in design and manufacture have made many flooring materials easier to install and less taxing on the homeowner's skills. For example, ceramic tile, once thought to require the skills of a professional, can be installed today by anyone with patience and average building talent. Intricate mosaic patterns combine sheets of tiles with a backing of plastic, cotton mesh, or paper that makes the tile easy to handle. Hardwood flooring can be purchased with durable factory-applied finishes. New types of carpeting materials and simpler methods of installation make laying some carpeting a relatively easy job for today's homeowner; seams in tufted carpeting can be ironed together with hot-melt seaming tape. Vinyl tiles come with adhesive backing or can be set in easy-to-apply adhesives.

It's a good idea to take the time to visit flooring dealers, home improvement centers, and flooring material suppliers to see the materials available in your area. Most dealers are happy to let you borrow samples so you can see how they look in your home. Tile suppliers will also help you choose the proper adhesives, grouts, and tools. Once you're familiar with the range of possibilities for flooring, you'll be able to narrow down your choices based on your needs and your own tastes. Coming to a final decision between wood, ceramic tile, resilient tile or sheet flooring, masonry or carpeting, is made easier by the following guidelines.

Aesthetics play a very important role in the decision-making process. The flooring material's design, texture, and the way it feels underfoot can set the mood of a room, even of an entire house. An area carpeted in a low, thick plush becomes a warm and inviting gathering place. A room laid with bold masonry blocks can appear to bring the outdoors into the house. Make sure that the flooring you choose communicates the feeling of the room and complements its decor. Since good-quality flooring will last for many years—often as long as the house itself—avoid designs, colors, or patterns that may date quickly.

Determine the kind and amount of traffic the flooring will have to bear. For high-traffic areas, such as hallways, entryways, kitchens, and bathrooms, select the most durable materials you can find. Use less rugged materials for areas that will receive less wear.

Cost is another factor. All the standard flooring materials come in various grades, with the cost directly related to the quality. Generally, the best-quality materials provide the longest and most satisfactory service, and are the most economical in the long run. A good rule of thumb is to install the best quality you can afford.

In terms of comfort, some flooring materials will be softer underfoot than others. If you object to a cold, hard surface, you may not want to use ceramic tile or masonry; wood or resilient flooring provides a less firm surface. Softest of all, of course, is carpeting.

If noise is a concern for you, realize that soft flooring materials, such as vinyl and carpeting, deaden sound. Wood, ceramic tile, masonry, and other hard surfaces tend to reflect sound, rather than absorb it.

Safety is an important factor to consider when choosing flooring. Keep in mind that although glazed tile is common in kitchens and bathrooms, it becomes slick when wet. Loose rugs laid on hardwood floors should have nonskid backings or pads.

Most types of flooring can be installed by a competent do-it-yourselfer. For instructions, turn to the chapters on wood flooring *(page 27)*; resilient flooring *(page 60)*; ceramic tile *(page 76)*; masonry *(page 88)*; and carpeting *(page 97)*. Check the installation instructions first to see if the flooring that interests you is within the range of your abilities.

New floor materials, protective finishes, and cleaning techniques make maintenance less of a consideration now than in the past. Still, flooring that's heavily traveled will get dirty and eventually may scratch or scuff. Care and maintenance instructions for the various flooring materials are in each installation chapter; take time to read them now so you'll know how much work you'll need to do to keep your new or refinished floor in mint condition.

Understanding the basic characteristics of the various materials and how they can be used will help you make your decision. On the following pages, we present brief descriptions of five basic flooring choices so you can compare the advantages and disadvantages of each.

 ASK A PRO

WHAT SHOULD I DO ABOUT MY SUBFLOOR AND UNDERLAYMENT?

Don't make any final flooring decision until you know the kind and condition of the subfloor and underlayment your new flooring will cover. With proper preparation, a concrete subfloor—because it's rigid—can serve as a base for almost any type of flooring. Other subfloors are more flexible and not suitable for rigid materials, such as masonry and ceramic tile. Refer to the installation sections of this book to determine not only the type of subfloor and underlayment required for each flooring material, but also the preparations and repairs you may need to make on your subfloor and underlayment before you can begin. It's a good idea to check the requirements with your flooring manufacturer.

WOOD FLOORING

Despite the development of far less expensive synthetics, wood flooring remains as popular today as it was in the past. Floors made from wood are usually called hardwood floors, even though they're sometimes made from softwoods such as fir, pine, hemlock, and redwood. Commonly used hardwoods include oak, maple, birch, and beech.

Wood floors that are properly sealed resist stains, scuffs, and scratches. When the floor is worn, it can be refinished to look like new. High-grade hardwood is expensive, but the cost is somewhat offset by its durability. Wood floors may shrink in heat or swell in dampness. They require a very carefully prepared subfloor and a moisture-free environment.

Almost all wood flooring can be classified into three basic types—strip, plank, and parquet. Each has its own character and special uses.

Strip flooring is the most common wood flooring found in homes, and is composed of narrow tongue-and-groove boards laid in random lengths; the widths don't vary. Strip flooring is available with tongue-and-groove ends or butt ends.

Plank flooring has a close association with our heritage—wide hardwood and softwood boards were readily available in colonial days. Maple planks more than a foot wide and secured to the subfloor with pegs can still be found in New England homes built in the 18th and 19th centuries. Today, plank flooring—boards produced in varying widths and random lengths—is still available. Unlike the planks of old, boards are now commonly milled with tongue-and-groove edges; screws add extra security and are usually covered with decorative plugs.

Parquet flooring is produced in dozens of patterns, textures, and thicknesses, and is made from solid pieces of wood, laminates, or individual pieces of wood held together by a backing. Though usually manufactured in squares, parquet can also be rectangular. A parquet floor is easy to install, yet has the look of a custom-designed floor.

The grain in wood floors provides natural texture and pattern. Both strip and plank flooring have their own intrinsic design; often this design can be enhanced by the way in which the strips or planks are laid. You can create quite a different look by making a parquet floor from scrap lumber cut in end-grain blocks. Special effects in existing wood floors can be produced by use of durable paints, colorful stains, special finishes, or stencil patterns.

The warmth of oak-plank flooring is seen to advantage in this finely crafted home. Random length tongue-and-groove boards, blind-nailed with a floor nailer, assure a tight fit and hidden fastening. The staircase's cedar balusters and post set off the oak treads and handrail.
(Architect: Ben Tarcher)

Laid diagonally, grooved oak-plank flooring provides an arresting visual tension in this dining area. The wood's natural grain has been highlighted by the use of a dark stain. A single wide board, equal to the wall's thickness, marks the junction of two diagonals, helping to define the rooms.
(Design: Audrey M. Borland)

Lightly stained hemlock creates a delightfully casual floor for this room. These 2x6 tongue-and-groove boards have been installed upside down so that their V-grooved top sides form a ceiling for the room below.
(Architect: J. Alexander Riley)

Mayan cherry parquet tiles add an air of refined antiquity to this dining room.

Parallel rows of finger-block parquet create a visually expansive living area, and make a refreshing alternative to the more common checkerboard pattern. (Architect: John Brooks Boyd)

RESILIENT FLOORING

The development of resins and synthetics has created a family of floor coverings called resilient flooring. Linoleum, one of the earliest examples of this type, is made from ground wood, cork, linseed oil, and resins, but it has been largely superseded by newer polyurethane and vinyl materials that are flexible, moisture- and stain-resistant, easy to install, and simple to maintain.

One of the advantages of resilient flooring is the variety of colors, textures, patterns, and styles available; cost varies widely depending on the material. Resilient flooring is relatively soft, making it vulnerable to dents and tears caused by sharp objects and large appliances, but often such damage can be repaired.

Available in sheets up to 12 feet wide or in standard 12-inch-square tiles, resilient flooring is easy to handle, although installation still requires patience and care.

Sheet flooring, usually made of vinyl or vinyl and polyurethane, can cover an average-size room without seams; it's a practical choice for kitchens, bathrooms, laundry rooms, and other areas where water may be spilled.

Vinyl sheet flooring has a top layer of vinyl bonded to a backing (either felt or vinyl). When a soft layer is pressed between the top layer and the backing, it produces cushioned vinyl—comfortable to walk on, durable, and quiet.

Resilient tiles are made of cork, solid vinyl, or vinyl composition (a blend of vinyl and other dense materials). Rubber tiles are also available, but are mainly for commercial installation. Tiles are easy to install, but care must be taken to lay them squarely to avoid gaps between tiles.

It's easy to create personalized designs with resilient tiles because they come in a wide range of colors, and are easily cut. Innovative use of resilient tile involves little more than careful planning: Make a scale drawing on graph paper and sketch in the design you've chosen; use one square on the paper for each piece of tile. Fill in the squares with the appropriate colors, to determine how much of each color to order, and to serve as a plan to set tiles of the right color in their proper positions.

The easy care of vinyl no-wax tile makes it a sound choice for high-traffic areas such as this entry hall.

Custom-designed white squares and long, pointed, midnight-blue pickets dramatize the elegance of resilient vinyl tile. Each piece is set off by a deeply beveled edge. Although the cost may be higher than usual for resilient flooring, the outstanding result justifies the outlay. (Design: Audrey M. Borland)

Sheet vinyl is extremely resistant to spills and stains, making it a natural for use in kitchens.

With the range of resilient patterns and designs available today, it's possible to obtain the look of ceramic tile without the expense. In this bathroom, the resilient sheet flooring resembles mosaic-bordered ceramic tiles.

Sheet flooring can also impersonate wood plank flooring, giving an authentic look to this country-contemporary kitchen.

CERAMIC TILE

The versatility of ceramic tile makes it a suitable flooring material for any room of the house. You can choose from the warm earth tones of quarry and Mexican tile or the brilliant colors and glossy finishes of glazed tiles. Glazed tiles are available in a wide variety of patterns, and you can even create complex designs and pictures with tiny mosaics. In choosing tile, you're not restricted to the traditional squares; many other shapes are available, from hexagons to elaborate ogee tiles.

Ceramic tile is extremely durable and is easy to clean and maintain; its resistance to water and stains makes it ideal for the kitchen or bath. However, ceramic tile is very rigid underfoot, reflects noise, and is susceptible to cracking if the floor shifts or if hard objects are dropped on it. Keep in mind that tile, particularly glazed tile, can be slippery. It's worth looking into special nonslip fin-ishes, particularly if you're laying the tile outdoors. For outdoor use, porcelain tiles are available that are impervious to water, making them able to withstand thawing and freezing. Other, more porous, tiles can be sealed to protect them from the weather.

Traditionally laid in a thick bed of mortar, ceramic tile can now be installed much more simply using a thin layer of adhesive applied with a trowel. Although tile is not difficult to install, care and patience is required to achieve neat rows and cuts. Ceramic tile requires a very flat, rigid surface, meaning that wood board floors and damaged resilient tile will have to be covered with ply-wood before the tile can be laid. Adhesives have been devel-oped that permit installation of ceramic tile in an area with some moisture, making it an appropriate choice to cover a below-grade concrete slab.

Each of the glazed modular tiles used for this kitchen-sunroom floor is a simple multiple or division of every other. The effect is an attractive, seemingly random pattern. (Architects: Jacobson/ Silverstein/Winslow)

A line of small tiles forms a border, with bright corner touches, in this bathroom floor detail. This is a simple, yet distinctive, way to add emphasis with tile. To create it, white tiles were cut to make room for the color accent. (Architects: Abrams, Millikan & Kent of Berkeley)

The use of hexagonal Mexican pavers in this entry and living room carries the eye naturally into a welcoming interior. (Architect: Peter C. Rodi/Designbank)

Mexican pavers were a logical choice when this solar greenhouse was built. As a floor, they can take lots of abuse; as a passive-solar mass, they're ideal for soaking up the sun by day and releasing the warmth to the house by night. (Architect: Obie Bowman)

Unglazed quarry tile on this kitchen floor gives a clean, crisp look. At home both in the house and outdoors, unglazed tile is best coated with clear masonry sealer for easy maintenance indoors. Although available in a variety of colors, shapes, and sizes, red-clay tiles are the most typical.

MASONRY FLOORING

Masonry floors are good-looking, easy to maintain, and virtually indestructible. Natural stone, such as slate, flagstone, marble, granite, and limestone, has been used as a flooring material for centuries. Natural stone includes rough-hewn shapes and uniform tiles. Both marble and slate are typically available as tiles that are relatively easy to install. Slate varies in color from dark blue to gray and green. Marble has a more traditional, formal look, but can be slippery. Because of its weight, natural stone can be expensive if it has to be shipped any distance; stone also has the disadvantages of being quite noisy, as well as cold underfoot.

Manufactured masonry products, specifically brick and concrete, were once considered only outdoor materials but are becoming increasingly popular for interior flooring. Brick is attractive and relatively inexpensive. Because it retains heat, brick is suitable for use in passive-solar heating designs. Concrete is, of course, used for slabs and is a perfect subfloor for almost any type of flooring. But concrete can also serve as a permanent floor that is durable and waterproof. Exposed aggregate relieves the monotony of plain concrete and adds texture and traction. Other special treatments can give concrete character and color.

The weight of masonry materials is such that a very strong, well-supported subfloor is required. Bricks are made in full thicknesses and in "splits" (half as thick). Concrete slabs can take the weight of full-thickness bricks, but wooden subfloors require the lighter weight splits. If you're considering a masonry floor, particularly one of natural stone, check your local building code for subfloor requirements.

Masonry flooring is laid in a variety of ways. The chapter on masonry floors covers materials that require a thick mortar base: rough-hewn stone, and brick. Stones that come cut into pieces of uniform thickness and shape, such as slate and marble, can be installed with adhesive; follow the instructions in the chapter on ceramic tile. A concrete floor requires building forms and casting concrete, techniques that do not fall within the scope of this book; consult a professional.

Laid in mortar over a concrete slab, roughly trimmed stones lend solidity to this entry hall. Their installation took care and patience, but the final result more than justifies the effort. (Design: MLA/Architects)

This sunroom was created by dividing off part of an existing patio. The original brick patio makes an attractive indoor floor and unifies the two areas. (Landscape architect: Woody Dike)

One neat route to an attractive woodstove hearth is a slate panel that sits directly on the subfloor. The slate was selected for grain and color pattern and for an exact match to the thickness of the surrounding oak. (Design: Russ Williams)

Marble tile's old-world elegance is perfectly suited to the sophisticated furnishings in this entry hall.

The combination of exposed aggregate and redwood headers gives a strikingly contemporary feel to this floor. The floor was poured early in house construction; time and use allowed the redwood to acquire a weathered patina. Later, it was simple to scrub the floor and apply a sealer. (Architect: J. Alexander Riley)

CARPETING

Like resilient flooring, carpeting is available in an array of colors, styles, and materials, with prices that vary widely. Comfort and appearance are two characteristics that have long made carpeting a commonplace flooring option for living rooms and bedrooms. Increasingly, the durable, easy-to-maintain carpeting that is commonly used in offices, factories, and public buildings is now finding its way into residential use in family rooms and hallways.

Conventional carpeting requires a pad, or cushion, underneath. You can install it directly over a well-prepared subfloor or over any type of old flooring that's clean, smooth, and free from moisture.

Although natural fibers are still popular, synthetics dominate the market because of their lower prices and easier maintenance. Carpeting is available in wool, nylon, polypropylene (olefin), polyester, and acrylic, as well as combinations of these materials. Olefin is used in the manufacture of Berber carpets.

Half a dozen basic manufacturing processes are used to produce carpeting. Among the styles that result are plush, level loop, and sculptured pile.

Plush carpeting has a smooth, even surface that gives the carpet a soft look. Because the pile is closely woven, the material doesn't crush easily. After vacuuming, it will have a shadowed appearance, although a type of plush carpet called textured saxony eliminates this problem by having the pile woven in different directions.

Level loop carpeting is made of tightly constructed loops that make it easy to maintain and very durable. Berber carpets come under this category.

Sculptured pile carpeting uses different lengths of pile to achieve varied surface levels. Rich surface texture and design characterize this type of carpeting.

Plush carpeting's smooth, level surface makes it an obvious choice for covering a large area like this dining room. Choosing a color is a very personal matter, but modern stain- and soil-resistant carpeting means that you can use light colors even in heavy-traffic areas.

This carpeting is enriched by a "sculptured" design. Different loop heights increase the carpet's durability and make possible this type of pattern.

Nubby, level-loop carpets like this one are loosely termed "Berber" carpeting. Berber carpets were traditionally made of wool, which has now been largely replaced by lower-cost synthetics such as olefin. Earthy Berber designs are equally good with natural and bright-hued color schemes.

This playful children's carpet is comfortable, practical, and educational. Good-quality carpeting not only insulates your home and absorbs sound, but advances in color and loom technology mean that almost any design or color combination is possible.

The beauty and distinctive texture of sisal are reproduced in this nylon multilevel loop carpeting. The natural color and basketweave motif can accommodate almost any style of interior decor.

BEFORE YOU BEGIN

Although many kinds of flooring are now designed for easy installation by the do-it-yourselfer, some projects are best left to professionals. If you intend to install your own, choose flooring that is not only suitable for the room where you plan to put it, but is at your skill level. Preparing a suitable base for new flooring can be more complicated and time-consuming than laying the final floor. The material determines the amount of preparation necessary; for example, ceramic tile, due to its inflexibility, requires a particularly stable subfloor. Keep in mind that not all types of flooring are suitable for all kinds of conditions; wood, for example, is affected by moisture, so only laminated products should be considered for rooms below grade.

If you're planning to do the work yourself, you'll want to know exactly what's involved in laying the particular kind of flooring that you've chosen. But before beginning, make sure that the floor is in good condition. Surface problems such as cracked ceramic tiles, loosened or cracked resilient flooring, or squeaks in floors and subfloors can often be traced to defects in the subfloor, joists, girders, posts, footings, or even foundation sills and wall studs. Structural problems caused by stress or deterioration in the foundation call for the expertise of a professional.

Before you undertake the installation of any new flooring—preferably before you even order the materials—take time to check the structure of the floor to make sure it's in good condition. The illustration below shows a floor and its supporting structure for a typical platform-framed house.

A few minor dips in floors, particularly in older homes, are not uncommon. Over a period of years, some settling results from stresses and fatigue. But if floors sag or are noticeably springy, you could have a serious structural problem.

To check your floor for sagging, take a straightedge at least 8 feet long and check systematically for the low spot as shown on page 4. A straightedge can be a length of rigid pipe; a piece of lumber at least 4 inches wide, selected for straightness; or the uncut edge of a strip of $3/4$-inch plywood. Mark the low spot, and, with a tape measure, "map" it in relation to two or more reference points (corners, ducts, or pipes) that you can use to locate it on the underside of the floor from the basement or crawl space.

Once you've isolated the location of the sagging, you can make a thorough inspection of the supporting structure. Joists should be checked for sway, sagging, or warping. Extreme softness in the wood is evidence of rot. Discolored wood on joists or on the subfloor usually indicates a plumbing leak. If the low spot is over or close to a post and the joists and beams are in good shape, the post or its pier or footing may be in bad condition. For any of these structural problems, call a qualified contractor.

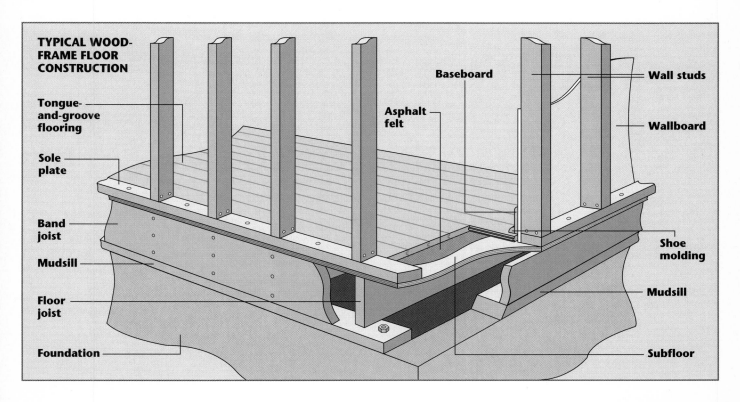

TYPICAL WOOD-FRAME FLOOR CONSTRUCTION

Tongue-and-groove flooring

Sole plate

Band joist

Mudsill

Floor joist

Foundation

Baseboard

Asphalt felt

Wall studs

Wallboard

Shoe molding

Mudsill

Subfloor

QUIETING SQUEAKY FLOORS

Squeaking floors have a number of causes: joists that are undersized or weakened by rot or termites; dried-out joists that have pulled away from the subfloor; inadequate nailing of the subfloor to joists; settling of the house; weak or inadequate bridging between joists; poorly manufactured wood strip flooring; warped floorboards; or sleepers that have worked loose from the concrete slab.

Squeaks in floors with finished wood surfaces can originate in surface areas or in the subfloor. In floors with flooring materials other than wood laid directly over the subfloor, the squeaks will come from the subfloor. In a typical platform-framed home with exposed joists (visible from the basement or crawl space), it's easy to locate the area of the offensive squeaks; this allows you to work from below and makes it unnecessary to nail through the surface of the floor. Solving the problem is more complicated in homes where joists are not exposed or when it's the upper floors that are making noise.

If joists are exposed, watch from below while another person walks across the floor above; you should be able to see the source of the squeak. Make sure the bridging between joists is firmly in place. If you see or feel movement in the joists, consult a professional.

Using a shim

TOOLKIT
• Hammer

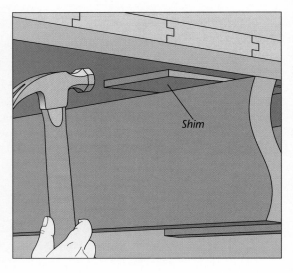

Shim

Tapping in the shim
For a minor problem, simple wood shims can be used to eliminate squeaks caused by movement between joists and the subfloor. Locate those areas under the floor where movement can be detected. Tap shims lightly into the gaps between the joists and the subfloor *(left)*, but take care not to drive them in too forcefully or they'll further separate the subfloor from the joists.

If the subfloor is made of individual boards laid diagonally, movement in an individual board or between boards can also be stopped with shims.

Installing a cleat

TOOLKIT
• Hammer

Nailing the cleat
Place a cleat—a length of 1x4 or 1x6 —against the joist and the subfloor; prop the cleat in place with a long piece of 2x4. Tap on the 2x4 to hold the upper edge of the cleat snugly against the subfloor *(right)*. Nail the cleat to the joist with 2" nails. No one should be standing on the floor above during the process.

If you observe movement between plywood subfloor panels in the space between joists, cut a 1x4 or 2x4 cleat to fit under the joint between the panels and attach it to both pieces of plywood with screws, thereby eliminating movement and stabilizing the panels.

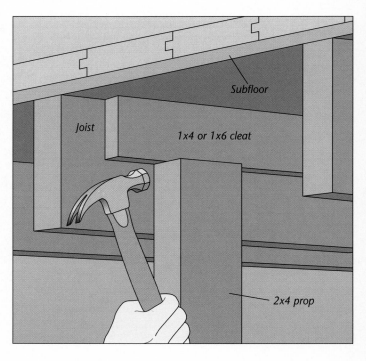

Subfloor

Joist

1x4 or 1x6 cleat

2x4 prop

TOOLKIT
- Drill
- Screwdriver or wrench

Using screws or screw devices

Pull loose boards tight by driving screws up through the subfloor. Use woodscrews in a length that will stop 1/4" short of the finished surface. Bore a pilot hole up through the subfloor with a bit the size of the screw's shank. Using another bit slightly smaller than the screw's threads, drill a pilot hole into the underside of the floor. Drill no closer to the surface than 1/4". Slip the screw through a large-diameter washer and up through the hole. As the screw is tightened *(below, left)*, the loose board will be pulled snugly against the subfloor. Instead of drilling pilot holes for woodscrews, you can drive drywall screws with a power screwdriver.

If it's the subfloor that's moving, you can use a commercially available device *(below, right)*; screw the plate into the subfloor, beside the joist. Hook the bracket over the joist, sliding it up the threaded rod. Use a wrench to tighten the nut until the subfloor meets the top of the joist.

Another method is to add L-braces *(page 26)*; when screwed onto the joist and then to the subfloor, these pull the floor down onto the joist.

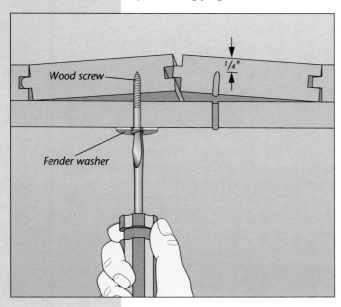

Wood screw · 1/4" · Fender washer

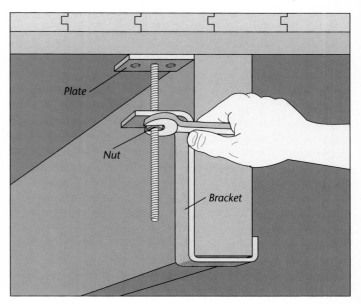

Plate · Nut · Bracket

Reseating loose nails

TOOLKIT
- Hammer

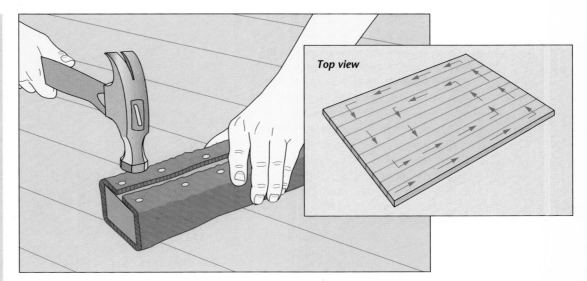

Top view

Using a padded block

If squeaking is limited to isolated areas, try to reseat loose nails. To avoid marring the floor, wrap a piece of 2x4 about 1' in length in an old towel, or tack carpeting to it, as shown above. Then place it over the squeaky area, at a right angle to the wood strips. Move the block in a rectangular pattern around the area *(inset)*; tap it sharply with a hammer. Avoid hammering over the same spot with too much force—this can damage tongue-and-groove flooring.

Several different kinds of lubricants that reduce friction between floorboards can be applied to the surface of the floor to eliminate squeaks. This is a useful remedy for squeaking floors that are inaccessible from below. Try the following possibilities:

Powdered or liquid graphite: Squirt powdered or liquid graphite between boards; it will work its way down into the tongue-and-groove joints. Use it very sparingly: Graphite is a messy substance, particularly if it is tracked across a floor.

Talcum powder: Dust cracks between boards with talcum powder and wipe up any excess with a damp cloth.

Floor oil: Apply a liberal coating of good floor oil over the squeaking area and wipe up any excess with a dry cloth. Oil that soaks down into the cracks will expand the wood and tighten the flooring to eliminate squeaking.

Mineral oil: If used in minimal quantities, mineral oil will help eliminate friction. A few small drops in the cracks between boards will be sufficient. Too much mineral oil can stain the surface of a floor.

Gluing the boards

TOOLKIT
• Putty knife

Inserting glue

If space permits, you can add glue to secure loose floorboards. However, most wood strip flooring has little or no space between boards. If your strip flooring has wide enough cracks between boards —space enough to insert a putty knife—common white glue worked into the cracks with the knife will help bind the boards.

After adding glue, push down hard and wipe away any excess. Put some weight on the glued area and leave it in place overnight. Books, bricks, or any heavy objects can be used, as long as you protect the floor's finish. Cover the glued area with a sheet of plastic before placing the weights to prevent gluing them to the floor.

Fastening the boards from above

TOOLKIT
• Drill
• Hammer
• Nailset
• Putty knife

Surface nailing

Surface nailing is a last resort. If working through a strip floor from above can't be avoided—because lubricants haven't worked or the floor is inaccessible from below—secure loose boards by nailing.

Drive ring-shank nails at a slight angle through the flooring and into the subfloor and, when possible, into a joist *(right)*. If the flooring is a particularly hard wood, drill pilot holes slightly smaller than the diameter of the nails. This reduces the risk of splitting boards and makes it easier to countersink nailheads. Drive and countersink the nails, select a wood putty that matches the finished floor, and fill the holes.

For a strip floor covered by carpeting, you can drive finishing nails through the carpeting. Avoid damaging carpet fibers with your hammer by using a nailset to drive nails home.

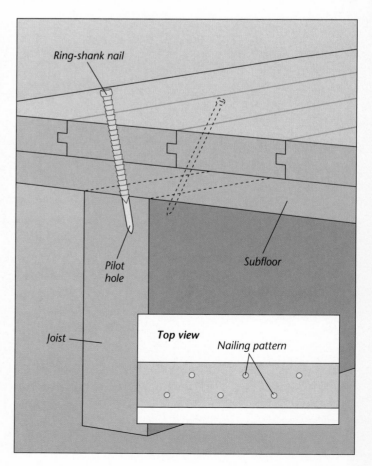

Ring-shank nail

Pilot hole

Subfloor

Joist

Top view

Nailing pattern

SILENCING SQUEAKY STAIRS

Probably more common than squeaks in floors, squeaks in stairs are caused by the same basic problem—wood rubbing against wood. The solution is either to eliminate movement between treads and risers or to lubricate squeaking areas.

Locating squeaks in stairs is easy. Move from step to step, pausing to rock back and forth on each step. Test the center of each tread first, then the ends.

If the undersides of the stairs are exposed—inside a closet, for example—eliminating squeaks can be relatively easy. When you don't have access from below the stairs, you'll have to work from above, which means that the repair may be visible. The same type of lubricants used on squeaky floors—graphite, talcum powder, floor oil,

or mineral oil—can also be used sparingly in the joints between the treads and risers of a staircase to eliminate friction. This solution can be applied only from above, and may prove to be temporary. However, since adding lubricant is a simple task, it can be repeated when necessary. (For details on using lubricants to eliminate squeaks, see page 24.)

Common white glue can also be worked into the joints between the treads and risers to bond them. This type of work is best done from above and is likely to be effective only where the movement between pieces of wood is minor. Work the glue into the crack with a putty knife or similar tool, and wipe away any excess glue immediately. Avoid walking on the step until the glue has dried.

Driving in wedges

TOOLKIT
• Pocketknife
• Butt chisel
• Hammer
• Utility knife

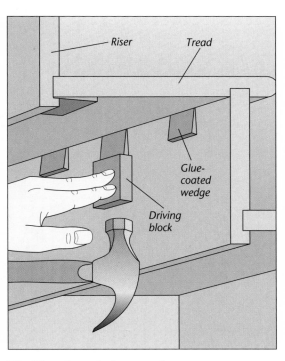

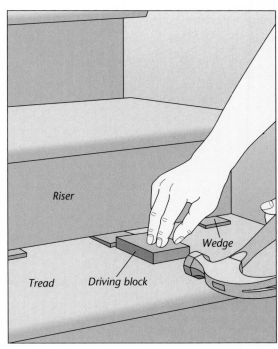

Working from below or above

Wedges are the simplest solution to squeaking stairs: Drive them between the riser and the tread, from below or above. While a helper puts weight on the squeaking step, look for any noticeable movement. Wherever there is play between the tread and the riser, you'll need to install wedges.

Wedges are best driven up from below, between a riser and the lip of the tread above it. The wedges should be small (1" or 2" long); you can whittle them off a board or shingle, using a pocketknife. Before you drive them into the riser, use a butt chisel to remove any shoe molding. Dip the tip of each wedge into white or yellow wood glue. Weight the step; then take a block of wood, place it against the blunt

end of the wedge, and tap the block to drive the wedge snugly into place *(above, left)*. Be careful not to pry the stair and riser apart; two or three wedges should be adequate. When the glue has dried, use a utility knife to cut off the protruding ends of the wedges, if you wish.

Where it's possible to work only from above the stairs, insert wedges between a tread and the riser above it *(above, right)*. Coat the tip of each wedge with glue and when driving a wedge, always use a block of wood to provide a larger hammering surface and to keep from marring the stairs. Carefully cut off the wedge's end with a utility knife when the glue has dried.

Using wood blocks

TOOLKIT
• Drill
• Screwdriver

Fastening wood blocks

Hardwood blocks work better than wedges where you have access to the underside of the stairs. Glue wood blocks (at least 1x1s) tightly to the tread and riser where they join, and secure the blocks with woodscrews in both directions. Installing the screws will be easier if you drill pilot holes in the blocks. Be sure to select screws that are not too long—they should come no closer than 1/4" to the surfaces of the treads and risers.

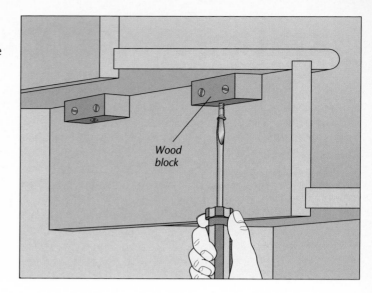

Wood block

Using shelf brackets or L-braces

TOOLKIT
• Screwdriver

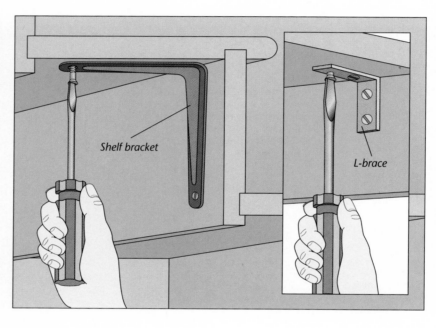

Shelf bracket

L-brace

Attaching the hardware

Shelf brackets or L-braces are easier to install than wood blocks, though not as effective. Install the bracket or brace so it is 1/4" below the top of the riser. When screwed in, it will pull the tread down onto the riser.

Choose short wood screws that come no closer than 1/4" to the surface of treads and risers when installed.

Installing molding

TOOLKIT
• Hammer
• Putty knife

Fastening the molding

When there's no alternative, quarter-round or any decorative molding makes a practical remedy that can be installed from above the stairs. Molding can be glued, then nailed to the angles between treads and risers to eliminate movement. The molding should be nailed to both surfaces.

Although you may need molding on only one or two stairs, you should install it on every step to give the stairway a uniform, finished appearance. For the same reason, all nail holes should be filled with putty, and the molding finished. Molding can be finished before it is cut and installed. This way, it only needs touching up once it has been nailed in place. NOTE: Check your building code to make sure the tread will still be long enough.

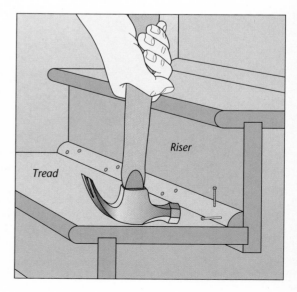

Riser

Tread

WOOD FLOORING

Wood is a traditional flooring material for North American homes and was widely used in bygone days because it was abundant and inexpensive. Good-quality wood flooring is no longer low in cost, yet wood remains popular because of its warm, natural look and its resiliency. Although the original cost is high, a good wood floor will last the lifetime of most homes, can be refinished several times, and will actually improve in appearance with age.

Wood flooring may be bought prefinished, or unfinished for sanding and finishing in place. Red and white oak are the most common species, but other hardwoods available include ash, cherry, maple, pecan, and more exotic species. Some softwoods, such as fir and pine, are also used.

In this chapter, we'll help you choose your wood flooring *(page 28)* as well as the tools and supplies for installing it *(page 30)*. Preparation of the subfloor is critical; turn to page 32 for information on making sure your floor stays flat and dry. Starting on page 38, you'll find instructions for installing and finishing three types of wood flooring: strips, planks, and blocks. Finally, we'll show you how to make repairs *(page 48)* and how to refinish a wood floor *(page 54)*.

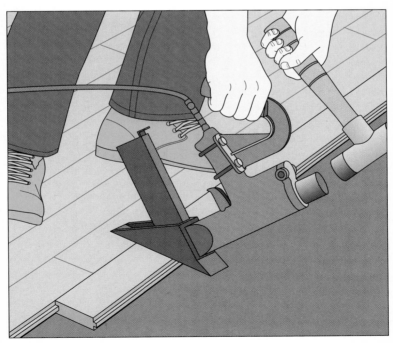

Tongue-and-groove strip flooring is nailed at an angle through the tongue into the subfloor or joists. A pneumatic nailer will make your work much faster and easier.

CHOOSING WOOD FLOORING

Wood flooring is generally not very difficult to install, but before you decide to work with it, consider the fact that preparing a base for wood flooring can be more demanding than putting in the new flooring itself. In fact, moisture can make some rooms unfit for wood flooring. In particular, wood floors are seldom installed in rooms that are below grade or in areas that are subject to dampness, unless laminated parquet or planks are used and careful steps are taken to keep moisture from reaching the wood. For details on preparing a subfloor for wood flooring, turn to page 32.

TYPES OF WOOD FLOORING

There are three basic types of wood floors: strip, plank, and block (known as parquet). The first two actually contribute to the structural strength of a house; most parquet flooring, on the other hand, is considered a floor covering only. A more recent option is known as "floating floors." This type of flooring is made up of tongue-and-groove boards that are glued together and laid unfastened over a layer of foam. For installation, follow the manufacturer's directions.

Strip and plank flooring are graded according to color, grain, and such imperfections as knots. Grading systems vary depending on the type of wood and the manufacturer. A typical system for oak designates the highest grade as Clear, followed by Select, No. 1 Common, and No. 2 Common.

Strip flooring: This is the basic type of hardwood flooring, made up of narrow tongue-and-groove boards that are laid in random lengths, with individual boards in varying lengths.

The most commonly used strip flooring for finishing in place is 3/4 inch thick, with a face width of 2 1/4 inches, although you can buy pieces with widths as narrow as 1 1/2 inches. This is suitable for most residential use. Thinner strips (5/16, 3/8, and 1/2 inch) can be used if there is a height restriction for the new floor.

Strip flooring is usually end-matched, with a tongue at one end and a groove at the other end of each piece. The boards also have a groove on the back to give the floor resiliency and to make it easier to lay boards over minor subfloor irregularities.

Plank flooring: A holdover from colonial days, most plank flooring sold today differs little from standard strip wood flooring. The major difference is that planks are produced in larger widths (3 inches and up), and you can order a mixture of widths. Plank flooring may be 3/4 or 3/8 inch thick and either solid wood or laminated—choose the latter for below-grade installations.

The individual boards are milled with tongue-and-groove edges, and square or tongue-and-groove ends. Because plank flooring can be screwed down at the ends, the boards may have holes already countersunk; wood plugs come with the boards. Plank flooring is available either prefinished or unfinished.

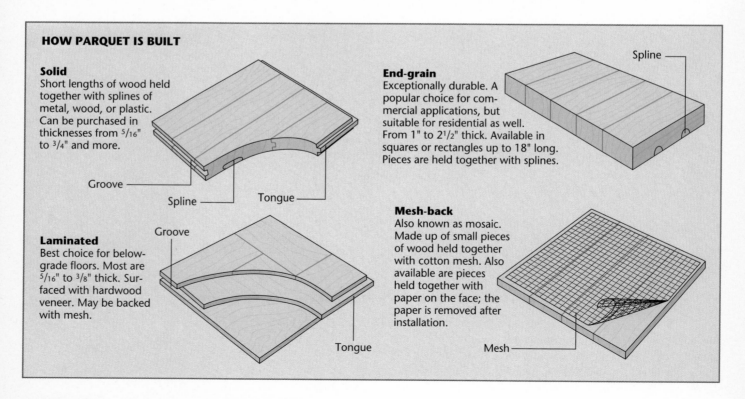

HOW PARQUET IS BUILT

Solid
Short lengths of wood held together with splines of metal, wood, or plastic. Can be purchased in thicknesses from 5/16" to 3/4" and more.

Groove — Spline — Tongue

Laminated
Best choice for below-grade floors. Most are 5/16" to 3/8" thick. Surfaced with hardwood veneer. May be backed with mesh.

Groove — Tongue

End-grain
Exceptionally durable. A popular choice for commercial applications, but suitable for residential as well. From 1" to 2 1/2" thick. Available in squares or rectangles up to 18" long. Pieces are held together with splines.

Spline

Mesh-back
Also known as mosaic. Made up of small pieces of wood held together with cotton mesh. Also available are pieces held together with paper on the face; the paper is removed after installation.

Mesh

Parquet flooring: The terms parquet, wood mosaic, and wood tile are used almost interchangeably to describe wood flooring laid in squares (or, sometimes, in rectangular blocks). Parquet flooring has become increasingly popular because of the wide selection of styles and the ease of installation. Parquet flooring comes in virtually every kind of wood and in a wide selection of patterns. You can buy it with a factory-applied finish, or, if you're ambitious, you can sand and finish it after installation.

As shown opposite, the squares may be solid wood, laminated sheets, or squares assembled from smaller wood pieces. Laminated squares are the best choice for floors that are below grade. Parquet flooring is most commonly made in 4- to 12-inch squares, but it's also available in rectangles or in panels up to 39 inches square and usually $5/16$ to $3/4$ inch thick, held together with mesh backing.

Parquet is usually applied with adhesive, much like ceramic tile. However, self-stick parquet is also available; although it's easier to install, it requires a very clean, dry, and smooth surface for it to adhere properly.

ORDERING AND STORING WOOD FLOORING

A flooring materials supplier will be able to tell you the quantity and cost of your flooring if you provide the exact measurements of the area, or preferably, a scale drawing. Order enough flooring to allow for waste and to leave extra pieces for future repairs.

Wood flooring must adjust to conditions in the room where it's to be installed, so plan ahead. When you order, discuss the date of delivery so you'll be ready to receive the materials and store them properly.

Your flooring should be delivered properly dried. If it's allowed to absorb moisture at the building site, it will expand; then, when it's installed in a warm, dry room, it will shrink and crack. So wood flooring should never be delivered in rain or snow unless well protected, and it should not be stored outside or in a cold or damp building.

For a new house or room addition, have all doors and windows in place and the structure closed in before the flooring is delivered; wet plaster or masonry, which releases moisture, should be allowed to dry thoroughly.

Storing wood flooring

Stacking strips and planks

New, unwrapped strip or plank flooring should be delivered at least four days before installation. If possible, store it in the room in which it will be installed—or, in a room with a similar environment. The temperature and humidity in the room should be close to normal levels. Untie the bundles and stack the individual boards loosely *(below)* so air can circulate around them.

Wrapped flooring should be unwrapped to allow it to adjust to the room before installation.

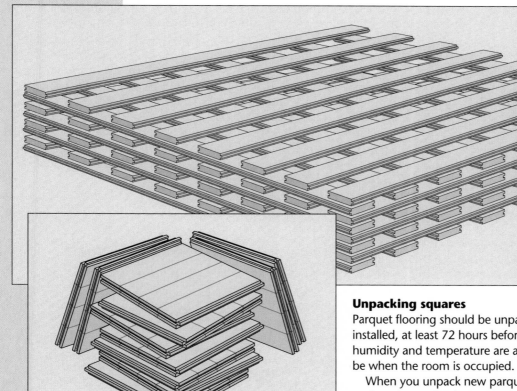

Unpacking squares

Parquet flooring should be unpacked and stored where it's to be installed, at least 72 hours before installation, but not until the humidity and temperature are approximately the same as they'll be when the room is occupied.

When you unpack new parquet flooring, stack it loosely *(left)* so that air can circulate around the pieces. Some parquet, particularly the prefinished type, is packed in sealed cartons; check the manufacturer's instructions before unpacking it.

TOOLS AND SUPPLIES

The tools and supplies you'll need will vary with the type of wood flooring you plan to install. Shown below are the tools you'll need to lay out your project. On the opposite page are the tools you'll need for actual installation. Some key tools, as well as the special supplies, are discussed below. You may need additional tools and materials to prepare a suitable subfloor, so before making your shopping list, read the information on preparing the subfloor, beginning on page 32.

If you install unfinished flooring, your dealer can recommend the appropriate sealer and finish. These products are discussed on page 54. Tools needed for finishing and refinishing are also described on that page.

Before you begin, be sure to have basic safety equipment on hand. This should include: work gloves for handling sharp or rough materials; eye protection for working with any striking tool; and rubber gloves to protect skin when handling harmful products. When operating any power tool, make sure to study the operating and safety instructions in the owner's manual. When installing parquet, you may find that knee pads make your work more comfortable.

Strip flooring (or unpegged tongue-and-groove planks): You can do a perfectly acceptable installation using basic hand tools. But the key to speeding up the task is to use a special nailer or stapler, available from most tool rental companies. Cleats (a special type of nail) or staples feed automatically into the tool and are then driven at the correct angle down through the tongue of each board to secure the floor. The mechanism is triggered by striking it with a rubber mallet. The tool is not difficult to operate and, after a little practice on a few pieces of scrap flooring, you should be able to master it with little trouble. A pneumatic nailer or stapler will make work even easier. Square-edged flooring is face-nailed by hand.

You'll need a supply of nails for blind-nailing tongue-and-groove boards. Even if you're using a nailer or stapler, you'll need a supply to nail those areas that can only be reached by hand. The nails are sized according to the thickness of the flooring—$2^{1}/4$- or $2^{1}/2$-inch ring-shank or cement-coated nails for $3/4$-inch-thick material, $1^{3}/4$-inch ring-shank or casing nails for $1/2$-inch-thick material, and $1^{1}/2$-inch casing nails for $3/16$- or $3/8$-inch-thick material. Your flooring supplier can help you select the right nails for the flooring that you buy.

You'll have to face-nail some of the boards and set the nails, so to fill these holes you'll need some wood putty that matches the color of the boards.

Pegged plank flooring: For maximum security, planks should be screwed as well as nailed. If the planks you buy don't already have holes drilled for screws and plugs, you'll have to drill these holes. The preferred method is to counterbore the plug holes with a Forstner or brad-point bit, then drill the countersink and the clearance and pilot holes for the screws with a combination bit for the size screw you'll use. If you're very careful to drill straight and hold the drill steady, you can drill the counterbores with a spade bit. You can also use four individual bits to make the counterbore, countersink, and clearance and pilot holes, but each hole will require four separate operations instead of only two. If you can borrow or rent a second electric drill, you'll be able to save time by not having to change bits frequently.

Depending on the thickness of the planks you'll be installing, you'll need No. 6, 9, or 12 flathead woodscrews; the thickness of the planks and subfloor will determine the length of the screws. Your flooring materials supplier will be able to recommend the correct size screw for the planks that you've purchased.

You'll need either precut wood plugs—often supplied with plank flooring—or hardwood dowels from which to cut your own pegs. Plugs in a contrasting wood will create a striking effect. You'll also need common white glue for fastening the plugs.

Parquet flooring: To install parquet squares, the flooring manufacturer will suggest adhesives that will work well with your type of flooring. The kind of adhesive may also vary with the type of subfloor to be covered. Solvent-base adhesives are being taken off the market and should be avoided because of their toxic fumes. Water-base adhesive should be used instead, but make sure the adhesive you buy is intended for wood flooring.

Some manufacturers recommend that you use a heavy floor roller to seat the parquet flooring in adhesive. A floor roller should only be used with unfinished parquet. To avoid damage to the floor's surface, be sure the roller is covered with a resilient material that can be kept clean.

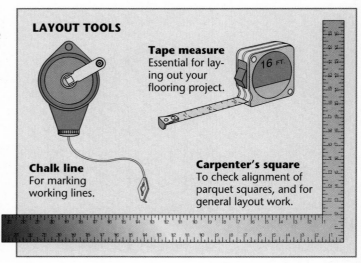

LAYOUT TOOLS

Tape measure Essential for laying out your flooring project.

Chalk line For marking working lines.

Carpenter's square To check alignment of parquet squares, and for general layout work.

TOOLS OF THE TRADE: WOOD FLOORING

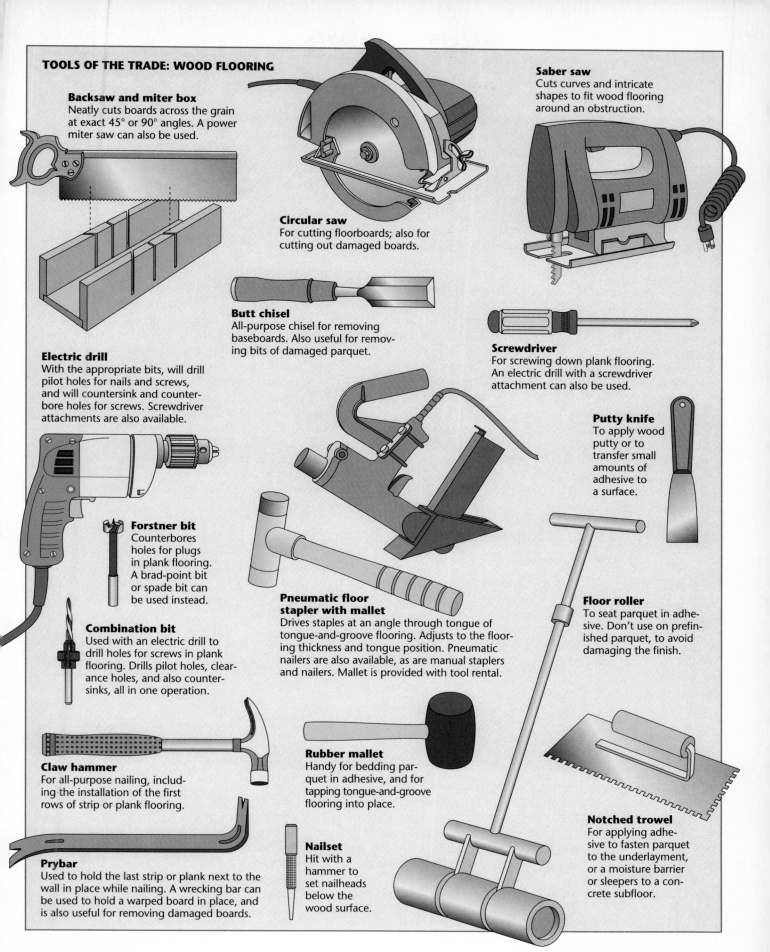

Backsaw and miter box
Neatly cuts boards across the grain at exact 45° or 90° angles. A power miter saw can also be used.

Circular saw
For cutting floorboards; also for cutting out damaged boards.

Saber saw
Cuts curves and intricate shapes to fit wood flooring around an obstruction.

Butt chisel
All-purpose chisel for removing baseboards. Also useful for removing bits of damaged parquet.

Screwdriver
For screwing down plank flooring. An electric drill with a screwdriver attachment can also be used.

Electric drill
With the appropriate bits, will drill pilot holes for nails and screws, and will countersink and counterbore holes for screws. Screwdriver attachments are also available.

Putty knife
To apply wood putty or to transfer small amounts of adhesive to a surface.

Forstner bit
Counterbores holes for plugs in plank flooring. A brad-point bit or spade bit can be used instead.

Pneumatic floor stapler with mallet
Drives staples at an angle through tongue of tongue-and-groove flooring. Adjusts to the flooring thickness and tongue position. Pneumatic nailers are also available, as are manual staplers and nailers. Mallet is provided with tool rental.

Floor roller
To seat parquet in adhesive. Don't use on prefinished parquet, to avoid damaging the finish.

Combination bit
Used with an electric drill to drill holes for screws in plank flooring. Drills pilot holes, clearance holes, and also countersinks, all in one operation.

Claw hammer
For all-purpose nailing, including the installation of the first rows of strip or plank flooring.

Rubber mallet
Handy for bedding parquet in adhesive, and for tapping tongue-and-groove flooring into place.

Notched trowel
For applying adhesive to fasten parquet to the underlayment, or a moisture barrier or sleepers to a concrete subfloor.

Nailset
Hit with a hammer to set nailheads below the wood surface.

Prybar
Used to hold the last strip or plank next to the wall in place while nailing. A wrecking bar can be used to hold a warped board in place, and is also useful for removing damaged boards.

SURFACE PREPARATION

Before you begin to prepare your subfloor, check for structural problems, as described on page 21. These problems must be solved before you proceed.

Wood floors are typically laid over a concrete slab, over a wood subfloor, or, in some cases, over an existing floor. Strips and planks can be fastened to a wood floor or subfloor, or to plywood or sleepers installed over a concrete slab. Parquet flooring requires a solid, smooth, continuous subfloor, whether of boards or plywood panels, concrete, or even resilient flooring. Laminated planks can be laid directly in mastic on a concrete slab. Solid wood should not be installed over concrete if there is any moisture problem; however, some laminated materials can be installed in this case.

CONCRETE SLABS

Except for glue-down parquet or laminated planks, wood flooring cannot be secured directly to a concrete slab. That means you'll have to lay either strips of wood (called sleepers) or a base of plywood over the slab so the flooring can be attached to the wood. Fastening plywood directly to the slab is a tricky job, but you may want to consider it if you're trying to limit the height taken up by the new floor. Sleepers will work only for strip or plank floors; parquet flooring requires a solid, continuous base of plywood—either over the slab or fastened to the sleepers.

Before installing any kind of wood floor over a concrete slab, you must make sure the slab is dry, level, and clean. A new concrete slab should be allowed to cure for at least 90 days before a floor is installed; make sure the area is well ventilated and turn the heat on during the cold season. New or old, it's essential that the slab be free from moisture year-round. Even if the slab appears dry, applying a moisture barrier under the floor will safeguard against future problems. If you are fastening plywood to the slab, you can first install a single layer of 4- to 6-mil polyethylene film, or a "two-membrane" system of asphalt felt. If you're installing sleepers, a polyethylene film will be laid over the sleepers.

If there is any evidence of a moisture problem, laminated flooring should be used with an adhesive also designed to serve as a moisture barrier.

Pay close attention to the specific guidelines for slab preparation and adhesives that accompany the flooring you're installing.

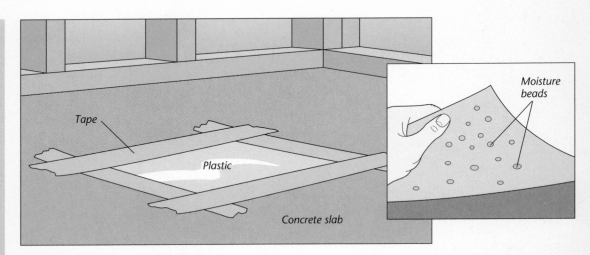

Preparing the slab

TOOLKIT
- Cold chisel
- Hand-drilling hammer
- Small trowel
- Straightedge
- Carpenter's level
- Notched trowel (for two-membrane moisture barrier)

1 Checking for moisture
In a damp climate, you may want to check the slab for moisture during wet times of the year. In any case, you should check all slabs for moisture not only from below but also from other sources—seepage through a wall or condensation dripping from a pipe.

To test if a slab is releasing moisture, cut a 1' square of clear plastic, tape it to the concrete floor (sealing the edges with tape), and leave it for two to three days. Then lift the plastic and see if any moisture has condensed on the other side *(inset)*. If there is only slight condensation, you can install parquet or laminated planks with a moisture-resistant adhesive. If the moisture problem is severe, check the downspouts, assess the drainage around the house, and look for leaky pipes. If the necessary repairs don't correct the problem, you should consider installing ceramic tile instead of wood.

2 Cleaning and leveling the slab

Once the slab is dry, make sure it's level and clean. Use a cold chisel to chip off mortar blobs or other spills of construction materials. Use a small trowel to fill minor dips with a patching compound or a leveling compound.

Sweep the slab clean to remove as much dirt and dust as possible. Do not use water to clean the slab; most auto supply stores sell a chemical cleaner that will remove grease and oil from a concrete surface.

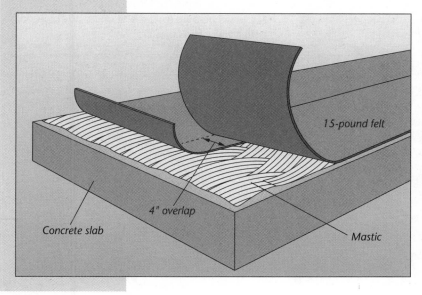

15-pound felt

4" overlap

Concrete slab

Mastic

3 Installing a moisture barrier

If you're fastening plywood to the slab the moisture barrier should be put down now. If you're installing polyethylene film, overlap the sheets 4" to 6".

To install a two-membrane barrier, use a notched trowel to spread a coat of asphalt mastic. Allow the mastic to dry. Then roll out a layer of 15-pound asphalt felt, overlapping the edges 4" and butting the ends of each course. Trowel on a second coat of mastic, then roll out a second layer of asphalt felt so that the seams run parallel to, but fall between, the seams of the first layer.

Note: A moisture barrier cannot prevent damage from standing water or flooding caused by improper drainage.

Installing plywood

TOOLKIT
- Circular saw
- Saber saw
- Ball-peen hammer
- Powder-actuated nailer (optional)

Fastening the plywood

To prepare a solid plywood nailing base over a concrete slab, first make sure that the slab is dry and clean, and install a moisture barrier as outlined above (step 3).

Lay out panels of 3/4" exterior-grade plywood over the entire floor, cutting the first sheet of every other run so that end joints will be staggered (right). Leave a space of 1/4" to 1/2" between panels and 3/4" around the walls. Around doors and other obstructions, where gaps will not be hidden with molding, cut the plywood to fit, leaving a gap of only about 1/8"; the flooring will hide these spaces.

Starting at the center of each panel, use concrete nails to secure the plywood to the slab. You may want to use a powder-actuated nailer. Use 32 nails per panel, in the pattern shown in the illustration. You'll attach your new wood flooring directly to this plywood base.

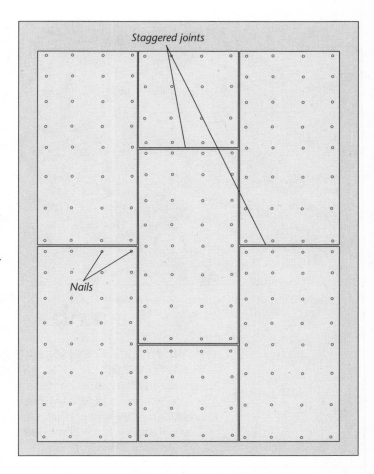

Staggered joints

Nails

Installing sleepers

TOOLKIT
- Notched trowel
- Straightedge
- Carpenter's level
- Plane
- Claw hammer (for installing plywood)

1 Laying the sleepers

The best material to use for sleepers is 2x4 lumber, pressure-treated with chemicals for pest- and moisture-resistance.

The simplest way to install sleepers is in a staggered pattern, as shown below. The 18" to 48" lengths are easy to handle; they also provide air circulation under the floor. Staggered sleepers are embedded in mastic on 12" to 16" centers. They're called "staggered" because the sleepers are overlapped at least 4" where they meet.

Sleepers should be installed at a right angle to the direction in which you're planning to lay your strip or plank flooring. To allow for air circulation under the flooring, no sleepers should be placed closer than 3/4" to any wall.

When the sleepers are in place, you may want to take a long straightedge (a straight 2x4 works well) as well as a carpenter's level and check to see that all the sleepers are level. Plane down any high points until they're level.

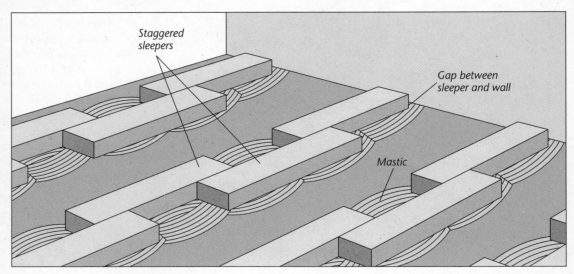

Staggered sleepers

Gap between sleeper and wall

Mastic

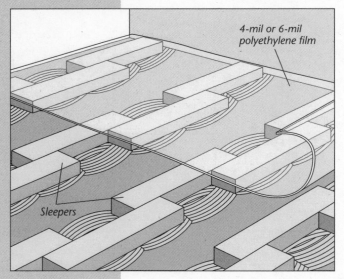

4-mil or 6-mil polyethylene film

Sleepers

2 Installing insulation, vapor barrier, and plywood

If you want additional moisture protection, a quieter floor, and some extra insulation, you can pour loose-fill insulation between the sleepers, leveling it with the top of the sleepers.

After leveling the sleepers, lay a 4- or 6-mil polyethylene film over the top of the sleepers for moisture protection. Overlap the edges of adjacent sheets of film at least 6"; it doesn't have to be attached. Take care not to puncture it.

You now have a good base for your strip or plank flooring. If installing a parquet floor, nail 3/4" exterior-grade plywood to the sleepers. Use 2" cement-coated or ring-shank nails and stagger the joints. Leave gaps between the plywood and the walls and doorways, as described for plywood applied directly to the slab *(page 33)* and 1/8" between panels.

PLAY IT SAFE

WORKING WITH PRESSURE-TREATED WOOD

Pressure-treated lumber contains potentially toxic chemicals; handle it with care. Most pressure-treated lumber available to the public uses inorganic arsenic (also called CCA, or waterborne preservative); only this type can be used indoors. Always cut pressure-treated wood outside; never burn it. Wear safety glasses and respiratory protection when cutting it.

NEW SUBFLOORS

A subfloor is installed over the joists. If you're putting wood flooring in a new home or addition—that is, if you can choose the sort of subfloor you'll be working with—install plywood; it is less expensive and easier to install than solid boards—and will provide a suitable nailing base for wood flooring. Because it won't warp, 3/4-inch plywood is generally considered the best subfloor material; it's also the easiest to work with.

Building a new subfloor

TOOLKIT
• Claw hammer
• Circular saw for plywood

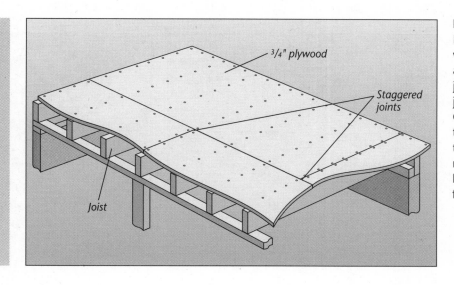

3/4" plywood

Staggered joints

Joist

Installing plywood
Lay plywood sheets with the grain running at right angles to the joists. Nail them to the joists about every 6". Center the ends over the joists and stagger the joints in adjacent runs *(left)*. Leave 1/8" between sheets to allow for expansion.

EXISTING FLOORS

If an existing wood floor is in good condition, you can use it as a base for new strip, plank, or parquet flooring; flooring can even be nailed through an old resilient floor. Parquet or glue-down plank flooring can be fastened to an old wood floor, to resilient flooring in good shape, or to plywood fastened over a resilient floor. If you are laying new flooring over old, you bypass the messy job of removing the old flooring and you gain some soundproofing and insulation. However, if you leave old flooring in place, there may not be enough space left above for appliances such as a dishwasher. And if you don't put new flooring under the appliances, you may find it impossible to remove them in the future.

Removing the old floor to expose the subfloor provides the most reliable base for your new floor and is a good opportunity to inspect the subfloor. Nail down loose panels; on a solid-board subfloor *(below)*, nail down loose boards. For best results, make sure you drive the nails into the joists. And if you're covering plywood, stagger the joints in the underlayment so they don't fall over those in the subfloor.

Repairing a solid-board subfloor

TOOLKIT
• Claw hammer
• Circular saw for plywood

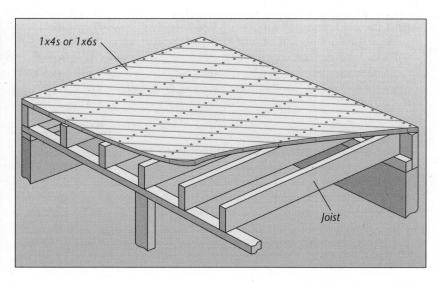

1x4s or 1x6s

Joist

Securing boards
To secure a loose board, drive two nails through it at every joist location, as shown at left. If the floorboards are too badly bowed to be flattened by nailing, give the floor a rough sanding with a drum sander *(page 55)*. If the subfloor is still uneven, cover it with 11/32" (or 3/8") underlayment-grade plywood.

TOOLKIT
- Butt chisel
  or prybar
- Claw hammer
- Nailset
- Drum sander
  (optional)
- Utility knife and
  floor scraper for
  removing resilient
  flooring

Removing trim and obstructions

Whatever kind of old flooring you're faced with, begin your preparations by removing all doors, grates, and other obstructions in the room. Use a chisel or thin-blade prybar to take up the shoe molding from around the baseboards *(below, left)*, prying it away every few feet. If the wood looks like it might split, find the nails and drive them through with a narrow nailset. Remove the baseboards in the same way *(below, right)*, taking care not to damage the walls or door frames. As you take up shoe molding or baseboards, number the pieces with chalk or a pencil so you can replace them in their original positions once your new flooring has been installed.

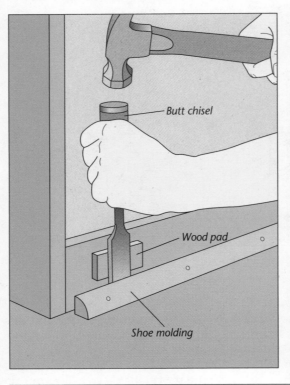

Butt chisel

Wood pad

Shoe molding

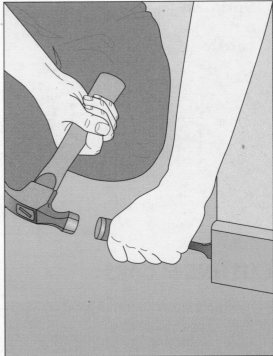

Preparing an old wood floor

An old wood floor must be structurally sound and perfectly level before you install new wood flooring over it. Separated, bowed, or buckling boards, or any evidence of moisture damage, serve as warnings of problems that you can't just cover over. If you have any reason to suspect the old floor or its supporting structure, remove the flooring and carefully inspect the subfloor, as described on page 21.

If you're satisfied the old floor is sound, set any exposed nails below the wood's surface, or pull them out. Secure loose boards by nailing them—directly into joists, where possible. Use ring-shank nails. Warped boards that can't be made level should be sanded flat or replaced, or you can cover the old floor with $^{11}/_{32}$" (or $^{3}/_{8}$") plywood as described above.

If you're planning to install new parquet flooring, you'll have to remove all wax, varnish, paint, or other finish from the old floor. This may require a rough sanding with a commercial drum sander *(page 55)*.

Removing or covering an old resilient floor

Old resilient flooring that is in good condition, flat, and securely fastened can serve as a base for new wood strips, planks, or parquet. If the resilient flooring is damaged, the floor should be removed or covered. CAUTION: Resilient flooring installed before 1986 may contain asbestos, which is hazardous if released into the air. Do not remove or sand it; cover it over. See page 72 for more information on asbestos flooring.

To remove old resilient tiles, see page 63. If the flooring is difficult to remove, it may be easier to cover it with $^{1}/_{4}$" plywood as described above.

FINAL FLOOR PREPARATION

Whether new or old, the subfloor (or old floor, if that is to serve as the base for new flooring) should be cleaned thoroughly. Drive protruding nails below the surface with a nailset and hammer, and correct any other irregularities to make a perfectly flat base. If you haven't already done so, you should remove the shoe molding from around the baseboards, as explained opposite.

If you've removed an old floor and are installing parquet flooring that is thinner than the old flooring, install plywood (or oriented-strand board, but check with your flooring manufacturer) directly over the subfloor to raise the surface of the new flooring to the same height as the old. This will ensure a level transition between the new floor and adjacent rooms.

Putting down asphalt felt

TOOLKIT
- Utility knife
- Straightedge or chalk line

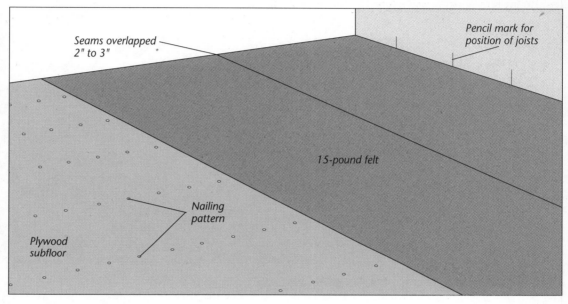

Seams overlapped 2" to 3"

Pencil mark for position of joists

15-pound felt

Nailing pattern

Plywood subfloor

Laying the asphalt felt

Unless you're laying the flooring in mastic, cover the subfloor with a layer of 15-pound asphalt felt (overlapping the seams 2" to 3"). Trim the felt flush with the walls using a utility knife. The felt will act as a moisture barrier, keep out dust, and help prevent squeaks in the new floor.

Flooring that is installed over the heating plant of a home needs extra protection to prevent it from drying and cracking. A double layer of 15-pound asphalt felt or a single layer of 30-pound felt will be effective. Standard $1/2$" nonflammable insulation board, cut to fit between joists from below, can be used as an alternative to heavier felt.

Before laying the felt, mark the position of the joists on the end walls, as shown above—the position can be established by the nailing pattern on the subfloor. These marks will serve as reference points when you start to fasten the new flooring.

ASK A PRO

HOW DO I PREPARE THE CRAWL SPACE UNDER THE NEW FLOORING?

If wood flooring is to be installed in a frame house with a crawl space, check the crawl space to see if a moisture barrier has been installed over the soil. If there is none, or if it has been torn or has deteriorated, use roofing felt or 4- or 6-mil polyethylene sheeting to cover the ground (overlap the edges by at least 2 inches). Use heavy objects such as stones or bricks to secure the moisture barrier in place.

You should also check to see that the crawl space is well ventilated. There should be at least four vents, one in each corner of the building, with a total vent area of about $1 1/2$% of the floor area—more if required by your local building code.

In dry climates, a moisture barrier may not be necessary, but providing adequate ventilation in the crawl space is still important.

INSTALLING STRIPS OR PLANKS

Before installing new flooring, prepare the subfloor *(page 32)* and give the wood time to adjust to the room humidity level. For new construction, all the walls should be in place. Both tongue-and-groove strip and plank flooring are attached by blind-nailing. With tongue-and-groove plank flooring, screws and plugs can be added—partly for strength, but mostly for the traditional look this gives. Some kinds of laminated planks can be glued down; these are installed much like parquet flooring *(page 43)*.

Install the flooring perpendicular to joists or sleepers. If you're nailing into joists, mark their locations on the wall, as described on page 37. For information on caring for your floor, turn to page 47.

Laying the strips or planks

TOOLKIT
- Tape measure
- Chalk line
- Electric drill
- Claw hammer
- Nailset
- Circular saw or power miter saw
- Rubber mallet (optional)
- Nailer or stapler
- Screwdriver (or screwdriver bit for electric drill) and center punch (for planks)
- Wrecking bar or lever-type can opener (for warped boards)
- Table saw for ripping boards (optional)

To frame around obstacles:
- Backsaw and miter box, or power miter saw

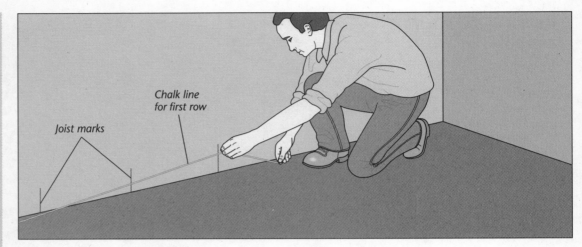

Chalk line for first row

Joist marks

1 Planning the first row

It's important to make sure that the first row you lay will be parallel to the center line of the room. Take several measurements of the width of the room and locate the center line as accurately as possible. Snap a chalk line to mark it. Next, measure from each end of the center line to your starting wall (place your tape measure perpendicular to the line), and snap another chalk line about ³/₄" from the wall you've selected as your starting point *(above)*; this will enable you to align the first row of flooring (with a gap for expansion). If you're working directly over sleepers *(page 34)*, using a chalk line will be more difficult, but the steps will be the same.

The space left at the edge of the flooring will compensate for minor irregularities in the wall. If you're starting from one wall of a room that's seriously out of square, rip some boards on an angle so that your first row of flooring will line up properly while keeping the proper distance from the wall.

 ASK A PRO

HOW DO I LAY OUT FLOORING IN AN IRREGULAR-SHAPED ROOM?

Establish a center line as close as possible to the center of the room and snap a chalk line. The first row of flooring will be laid from the center line, with the grooves in the boards toward the center. You'll then work toward the wall. When you return to the center, you'll need a special wood strip, or spline, to join two back-to-back grooved boards along the center line (right). If the walls are out of square, you may have to trim the last rows of boards.

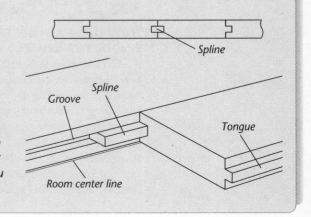

Spline

Groove

Spline

Tongue

Room center line

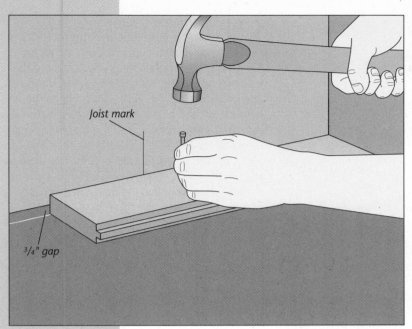

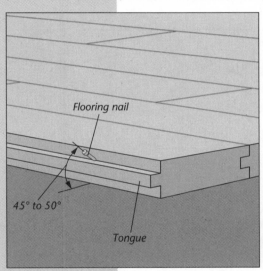

2 Nailing the first row

Work with one bundle of flooring at a time. Sort the boards by size where you're working; this will make it easier to plan the rows. Use the longest boards for the first row. If you're working with planks, start with the widest ones. Face-nail the first row near the wall where the nails will be covered with shoe molding; use the reference marks on the wall to nail directly into the joists *(left)*. Use 1½" nails. If you're working with especially hard hardwoods, drill holes in the boards slightly smaller than the diameter of the nails; this makes it easier to drive nails without splitting the boards.

If you're beginning at the center of an irregular-shaped room, you can blind-nail through the tongues of tongue-and-groove flooring or screw down square-edged planks, as explained in step 6.

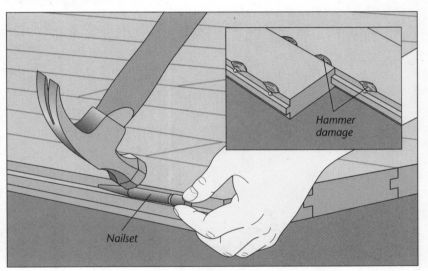

3 Nailing the next few rows

You won't have enough space to use a nailer until you are several rows from the wall, so you'll have to blind-nail the first courses of tongue-and-groove flooring by hand. Continue to drill pilot holes for the nails; this will help you keep the nails at the proper angle—45° to 50° to the floor *(above, left)*—and also prevent the boards from splitting.

Don't try to drive nails flush with your hammer; the indentations will show *(inset)*. Instead, leave each nailhead comfortably exposed; place a nailset sideways over it along the upper edge of the tongue, and drive the nail home by tapping the nailset with your hammer *(above, right)*. Then, with the tip of the nailset, set the nail flush.

4 Planning subsequent rows

Lay out boards six or seven rows ahead to plan the pattern. End joints should be staggered so that no joint is closer than 6" to a joint in an adjoining row. Find or cut pieces to fit at the end of each row, leaving a ½" gap between each end piece and the wall. As a rule, no end piece should be shorter than 8".

With planks, match widths of boards butted end-to-end. Work from the widest board to the next widest and so on; repeat the pattern. When working over sleepers, avoid having two adjacent joints fall between the same two sleepers. When working over plywood, avoid placing end joints over joints in the subfloor.

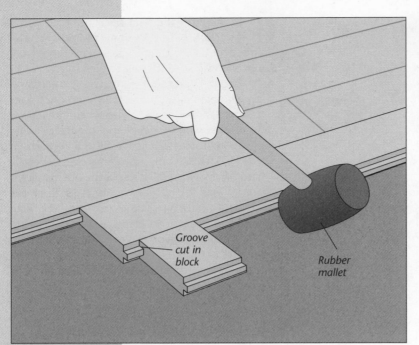

Groove cut in block

Rubber mallet

5 Tightening the rows

Each row of tongue-and-groove flooring should be driven tight against the previous one. Take a short piece of flooring, move it along the leading edge of the flooring you've just put down, and give it a sharp rap with a mallet or hammer before you drive a nail. If you're installing planks, note that some manufacturers of plank flooring recommend leaving a slight crack between boards—about the width of a putty knife blade—to allow for expansion in high humidity areas. Follow the recommendations of your flooring materials supplier for best results.

Periodically check the leading edge of the flooring as you work to make sure it's straight and still parallel to the center line.

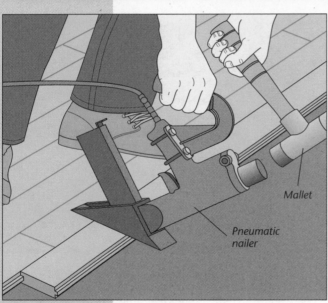

Mallet

Pneumatic nailer

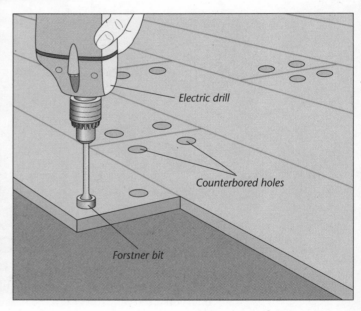

Electric drill

Counterbored holes

Forstner bit

6 Fastening the floorboards

Once you've nailed the first few rows, you can start to secure the flooring using the nailer or stapler. Adjust the tool to the correct angle—45° to 50° from the floor—and for the thickness of the flooring material. Position the tool, bracing the board with your foot; strike the tool with the mallet to drive the nail or staple home *(above, left)*. Use 2" nails or staples, unless you're nailing into plywood over a concrete slab—in this case, use a 1½" length.

Fasten the flooring at each end and every 10" along the board. Only the first and last rows (which are face-nailed) need to be nailed into the joists. When you're blind-nailing, the nails won't reach into the joists. If you're fastening into sleepers, drive nails or staples where the sleepers overlap.

To insert screws and plugs in planks, position the plank, insert the screws in the predrilled holes and tighten; an electric drill with a screwdriver attachment makes this easier. If the boards aren't predrilled, mark where you intend to drill, and use a center punch to tap in starter holes. Using the drill with the Forstner or brad-point bit, counterbore a hole ¼" deep *(above, right)*. Then drill a pilot hole, clearance hole, and countersink using a combination bit.

When the screws are in place, blow the dust out of each hole and fill the holes with wood plugs. Put a dab of common white glue on each plug before inserting it. Plugs should either be set flush with the floor surface or left slightly protruding (they'll be sanded flat when the floor is finished).

CAN I USE A WARPED BOARD?

Longer boards might have warped slightly because of moisture. You can still make use of one if you can hold it in place while nailing. You can use a block of wood and a wrecking bar (below, left). Nail the block to the floor, apply pressure with the wrecking bar, and then drive in a nail to hold the bar in place while you fasten the floorboard. Or, you can use a device made of 2x4 scraps, two strap hinges, and two lever-type can openers (below, right); step on the device while you fasten the floorboard—the can openers dig into the floor, pushing the board into place. Alternatively, you can cut the board at the spot where it is warped.

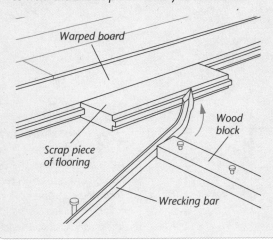

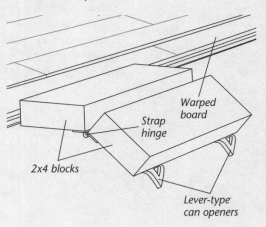

---

7 Installing the final rows

When you've progressed to the last few rows against the far wall (or from the center of an irregular room to either wall), you'll find it difficult to blind-nail the boards. Instead, drill holes and face-nail them using the reference lines on the wall to nail into the joists.

The final strip of flooring must be placed so that it will leave a 3/4" gap between the flooring and the wall. If you're lucky, a standard board will fit into the final strip. If not, you'll have to rip several boards down to the appropriate width, using a circular saw.

8 Framing obstacles

Exposed obstacles such as fireplaces should be framed to give a more finished look. Rip boards to fit between the flooring and the masonry. Leave a 1/4" gap and fill it with strips of cork available from your flooring dealer. If you're installing tongue-and-groove flooring, fit the framing pieces onto the tongues or grooves along the edges or ends of the floorboards. If a tongue will be left exposed on the framing, cut it off so it won't interfere with the masonry. You'll have to cut the corners of the framing at 45°. Use a miter box and backsaw to ensure an accurate cut *(page 42)*.

Floor openings covered with grates that have flanges *(inset)* don't need to be framed.

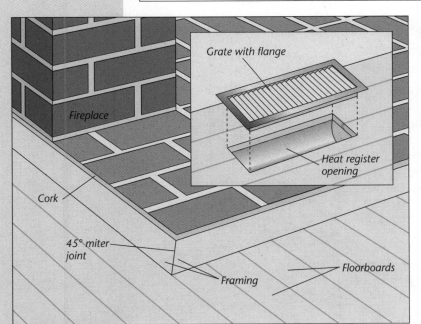

HOW CAN I MAKE PERFECT 45° CUTS?

For the corners on your framing to line up neatly, you'll need to make very precise miter cuts of 45°. The best approach is to use a miter box and backsaw. First, insert the saw: If you have a wooden miter box, simply choose the correct slot; if you have the type shown at right, set the degree scale. Align the cutting line with the teeth and hold or clamp the work firmly as you cut.

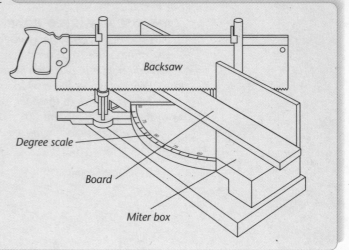

Backsaw

Degree scale

Board

Miter box

9 Installing a reducer strip

If your new floor will create a change of level from one room to the next, use a reducer strip for a smooth transition. A reducer strip is milled with a rounded top *(inset)*. It will fit onto the tongue of an adjacent board or, if laid perpendicular to the flooring pattern, onto the tongues at the board ends. It can also be butted against grooves' edges or ends. Face-nail the reducer strip at the floor's edge *(left)*.

Reducer strip

Side section view   Reducer strip

10 Reinstalling trim

If you've installed prefinished flooring, you can add (or replace) baseboards, shoe molding, and grates once the final board has been placed. Install shoe molding with a slight gap between it and the flooring; use a piece of thin cardboard for a spacer. Nail the shoe molding to the baseboard, not to the floor *(right)*, to let the floor expand or contract. The baseboard and shoe molding will hide the 3/4" gap between the last floorboard and the wall.

If you've installed unfinished flooring, you'll have to sand and finish the floor before you replace trim. For instructions on finishing your floor, turn to page 54.

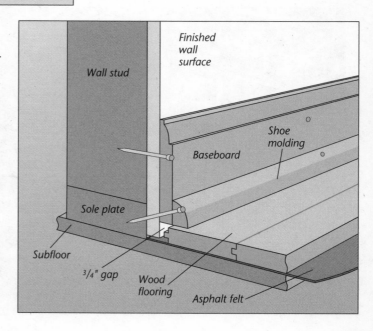

Finished wall surface

Wall stud

Shoe molding

Baseboard

Sole plate

Subfloor

3/4" gap

Wood flooring

Asphalt felt

INSTALLING PARQUET FLOORING

In this section, you'll find basic instructions for installing parquet flooring. Because parquet is produced in many different forms and can be finished in a variety of ways, follow the manufacturer's recommendations. A good parquet floor, properly installed, will have an exceptionally long life.

You may want to consider using self-stick parquet squares, but these require a very smooth, clean, dry surface and are not suitable for installation over concrete.

First, prepare a proper base for your parquet (*page 32*). Make sure that any walls are already in place before installing the flooring, and give the flooring time to adjust to the room's humidity level (*page 29*).

No matter how they may look, few rooms are perfectly symmetrical. You'll need to establish working lines independently of the wall lines, as described below. The following instructions show how to start your installation at the center of a room and work toward the walls. This is the best method to use if the room is badly out of square, or if you want cut tiles to be balanced along opposite walls. If the room is going to be viewed from one direction only, you may want to hide cut squares along one pair of adjoining walls; this can be achieved by starting from one wall. For instructions on this type of installation, turn to page 82. Keep in mind that there are no grout joints for parquet flooring.

Establishing working lines

TOOLKIT
- Tape measure
- Chalk line
- Carpenter's square

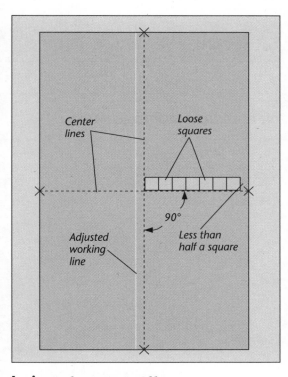

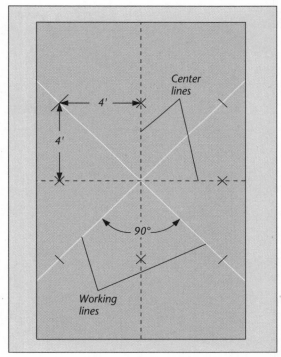

Laying out a square pattern

First, locate the center point on each of two opposite walls, and snap a chalk line across the floor between the two points. Then find the centers of the other two walls and stretch your chalk line at right angles to the first line, but snap the line only after you've used your carpenter's square to determine that the two lines cross at precise right angles.

Next, to ensure that the border units around the perimeter of the room will look balanced, lay a row of loose squares of flooring along each line from wall to wall. If the space between the last square and the wall is less than half a unit wide, move the center line the width of half a tile in either direction.

Laying out a diagonal pattern

If you prefer to lay your flooring in a diagonal pattern, establish working lines as described for the square pattern. Then mark each line at two points 4' from the center on either side. From these points, measure and mark the end of a 4' line in each direction, taking care to lay your tape measure exactly perpendicular to the center lines.

You should now have an 8' square, the center of which corresponds to the center of the room. Snap chalk lines between opposite corners of the square to get diagonal working lines. If your measurements were accurate, the diagonal lines should intersect exactly at a right angle over the center of the room. If they don't, check your measurements again.

Checking the working lines

Whether you've established square or diagonal working lines, you should make a final check to be sure the lines cross at an exact right angle. Measure 3' along one line and 4' along the other, then measure diagonally between these two points; this distance should be exactly 5'. If it isn't, adjust your working lines and recheck.

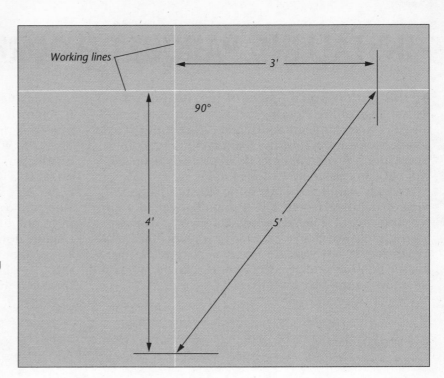

SEQUENCES FOR PLACING PARQUET SQUARES

Pattern A

11				
10	12			
4	9	13		
3	5	8	14	
1	2	6	7	15

Pattern B

Pattern C

Parquet should be placed in a pyramid fashion so that as many squares as possible will fit into corners formed by the squares that you've already laid. If you've established square working lines, you can fill in half the floor area at a time, as shown in pattern A; this method is the best choice if the open time of the adhesive that you're using allows you to cover a large area at a time. You can also fill in the floor area one quarter at a time, as shown in pattern B. If you've established diagonal working lines, use pattern C.

Laying parquet squares

TOOLKIT
- Putty knife
- Notched trowel
- Rubber mallet
- Carpenter's square
- Floor roller for unfinished floors
- Saber saw
- Crosscut saw
- Contour gauge
- Claw hammer

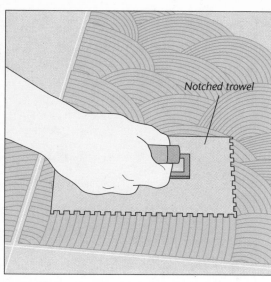

Notched trowel

1 Spreading the adhesive

Use a putty knife to transfer some of the adhesive from the container to the spot on the floor where you plan to lay the first block. Holding the notched trowel at a 45° angle, spread the adhesive evenly, putting firm pressure on the trowel (*left*). If you can't see your working lines between ribbons of adhesive, spread the adhesive up to, but not over, the chalk lines.

Check the open time of the adhesive (the time you have to lay the blocks before the adhesive has begun to dry). Don't spread more adhesive than you can cover in this time.

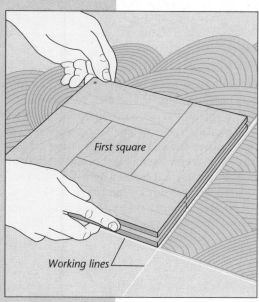

First square

Working lines

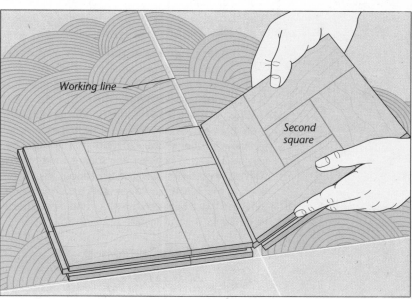

Working line

Second square

2 Placing the first squares

Align each square carefully; minor irregularities can become problems as you work across the floor. The illustrations that follow use placement pattern B (*opposite*). You can adapt these instructions to the other sequences shown, or to one suggested by the manufacturer. Place the first square at the intersection of the two working lines, aligning the edges (grooved edges, if tongue-and-groove) with the lines (*above, left*). Set the second square against the first. Align it carefully with the first one and with your working lines. Engage tongue-and-groove squares as you put them together (*above, right*). Don't slide the squares into place—this will force adhesive up into the joints.

3 Continuing the sequence

Place the squares in the pattern you've selected. Check your work frequently with a carpenter's square. After you've set several squares, tap them sharply in several places with a rubber mallet to seat them in the adhesive. If you're installing unfinished parquet, roll each section of the floor with a floor roller before the adhesive sets. (NOTE: This would damage prefinished parquet.) Any adhesive forced up between the squares or tracked onto the floor surface should be wiped off with a damp cloth. If the adhesive has dried, wipe it off with mineral spirits.

When you have to walk or kneel on the new flooring, place pieces of plywood over the squares to distribute your weight.

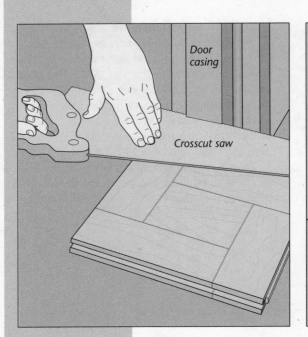

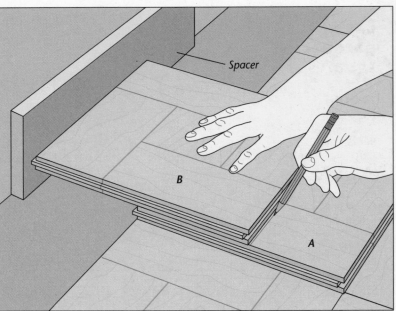

 Fitting squares around walls and doors
As you near the wall, you may need to trim a door casing so the square will fit under it. Use another square for a guide to make the proper cut.

When you reach a wall or other obstacle, it will probably be necessary to cut individual squares to fit. Allow for the natural expansion of the wood by leaving a 3/4" gap between the square and the wall; with other obstacles, leave a gap of about 1/16". Some flooring manufacturers supply cork expansion strips to fill the space between the block and the wall; use this material if it's available.

To cut a square to fit, mark it as shown above, right. Set the square to be cut (A) directly over a square that is already in place and then set another one (B) over it against the wall—use a pencil to mark where they overlap. A piece of scrap wood of the right thickness can be used as a spacer. Then use a crosscut or saber saw to trim the square. Cut and set the squares one at a time as you proceed. Where a square meets an irregular obstruction, you can mark where to cut the square by using a contour gauge, as explained below. Then use a saber saw to make the cut.

ASK A PRO

HOW DO I USE A CONTOUR GAUGE?
A contour gauge is used to copy an irregular shape, such as a door casing, onto a parquet square. Holding the contour gauge flat, push it against the obstruction. Pull it away—friction will hold the rods in place—and copy the shape onto the square using a pencil.

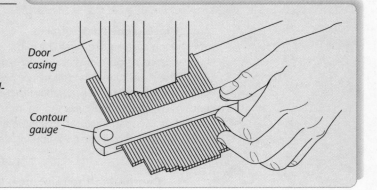

5 **Meeting old flooring**
If your new flooring will be level with the floor in an adjacent room, the new and old floors should meet under the door. The joint can be hidden with a wood saddle, available from your flooring supplier. If the new flooring is higher than the floor in an adjacent room, install a reducer strip *(page 42)* to adjust the level where the floors meet. This too is available from the flooring manufacturer in the same wood and finish as the parquet.

6 ▶ Reinstalling trim

Allow newly installed parquet squares to set overnight—or as long as recommended by the flooring manufacturer. If you've installed unfinished squares, you'll need to finish the floor before you replace the trim. Then replace any baseboards, shoe molding, and grates that were removed. Leave a slight gap between the shoe molding and the floor (use a thin piece of cardboard as a spacer). Nail the shoe molding to the baseboards, not to the floor. This will allow for the natural expansion of the flooring. If the floor level has been raised by the new flooring, you may have to trim material off the bottoms of doors before you rehang them.

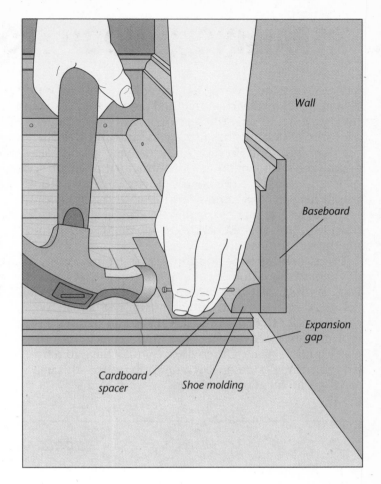

Wall

Baseboard

Expansion gap

Cardboard spacer

Shoe molding

CARING FOR YOUR WOOD FLOOR

Water is the natural enemy of wood. Because water seeping between boards can cause stains or swelling, avoid wet-mopping wood flooring and using water-base waxes. In general, you should vacuum or dry-mop wood flooring about once a week, or as often as you vacuum carpeting. Wipe up spills immediately with a damp cloth and dry immediately with another cloth.

Wood flooring is finished with either a penetrating sealer and wax or a surface finish. If you've installed a floor with a factory-applied finish, the manufacturer will specify which product has been used. However, if you don't know how your floor was finished, scratch the finish in a corner with a coin; if it flakes, it's probably a surface finish. Otherwise, assume a penetrating finish was used. A floor finished with a penetrating sealer may need to be waxed once or twice a year. Most manufacturers recommend that floors with a surface finish not be waxed because it will make the floor more difficult to refinish. If the floor finish—either penetrating or surface—is damaged in spots, it can be patched or renewed *(page 58)*. To refinish the entire floor, turn to page 54.

Cleaning floors finished with penetrating sealers: Do not damp-mop. For occasional cleanup of spots, use mineral spirits and fine steel wool. After wiping the area with a clean, soft rag, buff the floor lightly by hand.

When the floor shows signs of excessive wear, clean a small area at a time with a mineral-base cleaner, going over the area with rags or No. 1 or 2 steel wool. Wipe off any excess cleaner with a clean rag or clean steel wool. Then, finish with a thin coat of a good non-water-base paste wax (one intended for use on hardwood floors) and buff the floor to a satin shine.

Cleaning surface-finished floors: For periodic cleanup of spots and stains, go over the floor with a barely damp sponge mop. If any stains remain afterward, use a mild cleanser, such as white vinegar, to remove them. Rinse the floor with a clean, damp (but not wet) sponge mop to remove any residue and dry with clean rags. To restore luster to the floor, you can buff it lightly by hand.

Products such as mineral-base cleaners can be used for stubborn spots, but be sure to test the product on a hidden area first.

REPAIRING A WOOD FLOOR

If boards are badly warped, or there is evidence of water damage, there may be a structural problem—check your floor *(page 21)*. If only the surface of the floor is damaged, the best solution is usually to sand and refinish the floor *(page 54)*. However, if the damage extends deeper than the surface but there is no structural problem, you may want to replace the boards. It may be difficult to match new pieces and new finishes with the old floor.

A damaged square of parquet flooring is not difficult to replace; the toughest part of the job might well be finding the right substitute.

Tongue-and-groove flooring: It may be impossible to buy replacement flooring that perfectly matches your floor. To find a suitable replacement, take a sample to your flooring dealer. If your flooring is prefinished, a match may be easier to find. Sometimes, you can remove a replacement piece from a closet, then replace that with a less suitable piece. Otherwise, you may have to refinish the entire floor to give it a uniform finish—this will solve the problem of matching.

There are two common approaches to replacing boards in random-length hardwood floors. The first is to cut out a rectangle, remove the damaged boards, then replace them with boards of equal length *(below)*. This approach is satisfactory for areas that will be covered by a rug or furniture. The second and slightly more difficult method is to remove boards in a staggered pattern *(page 51)*. This produces a less noticeable repair and is best for an open floor area.

Parquet flooring: You may have a problem matching the finish of a replacement square with that of the old floor—especially if the floor was sanded and finished after installation. Reserve a piece of the damaged square you remove to show to a flooring dealer; this will help you find a match. Matching prefinished parquet flooring is somewhat easier, particularly if a few extra pieces were set aside at the time the original floor was installed. Even if you don't have any extras, you may be able to find a suitable match through a flooring dealer.

TOOLS AND SUPPLIES

For replacing floorboards with either the rectangular or staggered method, you'll need a few basic hand tools: a combination square; a sturdy, sharp 1-inch chisel; a claw hammer; and a wrecking bar. For cutting out a rectangle, you'll also need a circular saw. A portable electric drill with a spade bit will be useful for taking up individual boards if you're removing them in a staggered pattern.

To replace parquet, you'll also need adhesive. Choose a water-base adhesive—ask your dealer for one that will work well with the type of parquet you're using.

Replacing floor-boards: rectangular pattern

TOOLKIT
- Tape measure
- Carpenter's square
- Circular saw
- Claw hammer
- Butt chisel
- Wrecking bar
- Nailset
- Combination square
- Backsaw and miter box, or power miter saw
- Sanding block

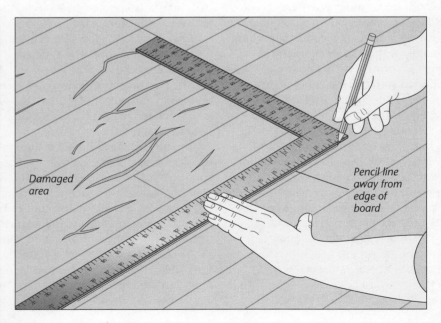

Damaged area

Pencil line away from edge of board

1 Marking the area

Mark the work area with a carpenter's square and pencil. Use a lengthwise joint to align the square for the end mark; make the side marks at least 1/4" away from joints so that the saw won't hit the nails.

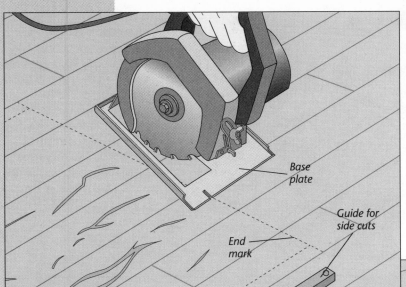

Base plate

Guide for side cuts

End mark

2 Making the cuts

Adjust the blade on a circular saw to cut nearly through the board. To make the end cuts, position the base plate of the saw at the midpoint of the end mark. With only the toe of the base plate resting on the surface *(left)*, turn on the saw; before moving ahead, slowly lower the blade until the base plate rests flat on the floor. Work from the center out to each edge. To guide the side cuts, measure the distance from the outer edge of the base plate to the blade and then tack a strip of wood at this distance from the pencil line. Make the cuts from the center toward the ends, keeping the edge of the base plate against the guide.

3 Completing the cuts

Using a hammer and chisel, complete the cuts through to the subfloor *(right)*. Keep the beveled face of the chisel toward the boards to be removed. Take care to make clean cuts at the corners, so that the boards don't splinter into the surrounding undamaged area.

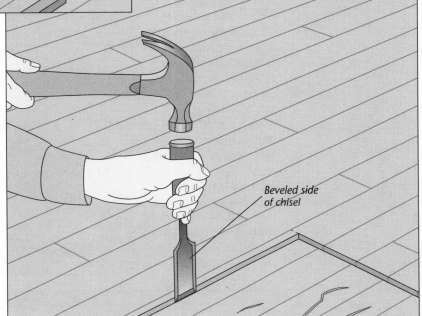

Beveled side of chisel

Wood block

Wrecking bar

4 Removing the boards

Lift the boards with a wrecking bar; use a small wood block for leverage *(left)* and to avoid marring any part of the surrounding flooring. Using a hammer and chisel, cut away the 1/4" edges left outside of the saw cuts, shown as dotted lines in the illustration. Work carefully to avoid damaging adjacent boards. Set any exposed nailheads with a nailset.

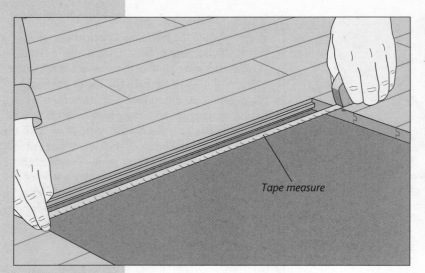

Tape measure

5 ▶ Cutting and fitting the replacement boards

You can usually mark the replacement board right in place. Cut one end at a precise 90° angle using a miter box and a backsaw. Fit the cut end tightly into the space and mark the other end for a tight fit. Use a carpenter's square to mark your cutting line and make the second cut to the waste side of the line. Repeat for additional replacements.

If the repair is close to a wall, you may need to cut the replacement piece roughly to size before marking it; or you can use a tape measure marked in $1/32$" or $1/100$".

6 ▶ Fastening the boards

Blind-nail the new boards at an angle through the tongues *(inset, below)*. Fit the groove on the new board onto the tongue of a board already in place. To avoid damaging the boards with the hammer, leave the nailheads exposed. Drive each nail home by placing a nailset over it sideways and striking with the hammer *(page 39)*. Finally, set the nail with the nailset *(right)*.

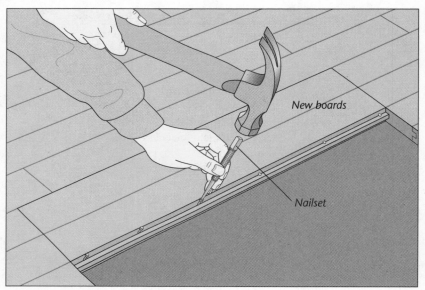

New boards

Nailset

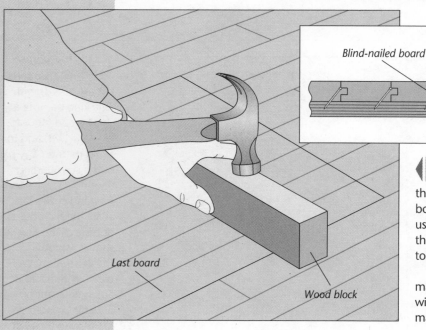

Last board

Wood block

Blind-nailed board   Face-nailed board   Tongue removed

7 ▶ Fitting the last board

Cut off the tongue and sand the cut edge so the board fits tightly; you may need to bevel the bottom. Set the board in place and tap it down using a hammer and a wood block *(left)*. Face-nail the board *(inset)*; set the nails. Or, remove the bottom part of the groove *(page 52, step 4)*.

Fill nail holes and any mismatched end joints with matching wood putty. Then sand the new boards with a sanding block; stain and finish them to match the surrounding floor *(page 58)*.

TOOLKIT
- Electric drill
- Butt chisel
- Claw hammer
- Prybar
- Backsaw and miter box, or power miter saw
- Rubber mallet
- Nailset
- Table saw (optional)
- Sanding block
- Putty knife

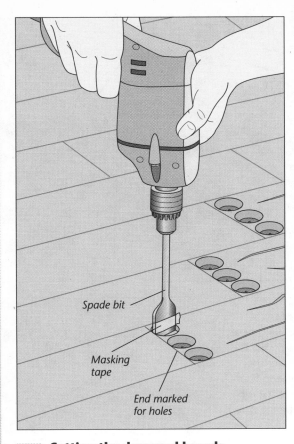

Spade bit

Masking tape

End marked for holes

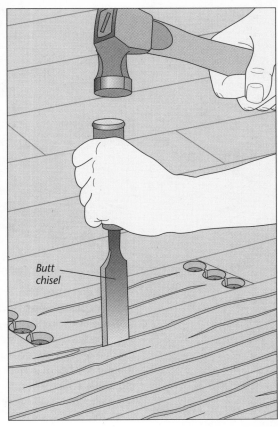

Butt chisel

1 Cutting the damaged boards

Mark the ends of board sections to be removed in a staggered pattern. Use a large spade bit to drill holes at both ends of the sections *(above, left)*. Use a masking-tape flag to indicate the drilling depth—don't drill into the subfloor with the bit's blade. Connect the holes with chisel cuts before splitting the boards or the splits may continue between the holes to undamaged boards.

Break up the damaged floorboards by splitting the damaged area of each board using a large butt chisel and a hammer *(above, right)*. Tap lightly to avoid splitting or cracking the surface of adjacent boards.

Prybar

2 Prying out the pieces

Pry out the split lengths of floorboard using a prybar *(left)*. Use a small block of wood for leverage (as shown on page 49, step 4), and to avoid marring any part of the surrounding flooring. Carefully trim the edges of the cut area square using a hammer and butt chisel.

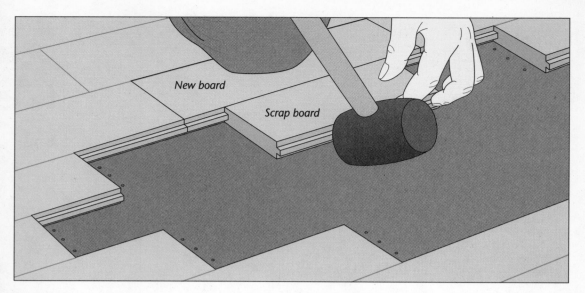

3 Fastening the replacement boards

Cut and fit the replacement lengths as described on page 50, step 5. Slip the groove of the new board over the tongue of a board in the existing flooring; use a scrap of flooring and a rubber mallet to tap the new board into place *(above)*. Blind-nail the new pieces tightly in place. If adjacent existing boards have separated, use thin shims *(right)* to align the edges of the new boards with the edges of the old ones.

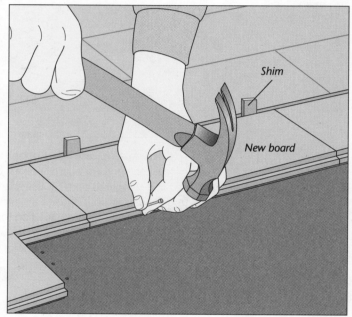

4 Installing the last board

To install the last board, you can cut off the tongue as described on page 50, step 7. Alternatively, you can remove the bottom lip of the grooved edge, as shown in the inset, with a hammer and chisel or a table saw. Slip the tongue into the groove of the existing board *(left)* and press the new board into place; face-nail both ends. Fill nail holes and any mismatched end joints with wood putty. Sand the new boards with a sanding block and stain and finish them to match the surrounding floor *(page 58)*.

Replacing parquet

TOOLKIT
- Circular saw
- Hammer
- Butt chisel
- Cold chisel (optional)
- Table saw (optional)
- Sanding block
- Notched trowel
- Rubber mallet

1 ▶ Removing a damaged parquet square

Set the blade depth of your circular saw to the thickness of the parquet square. Make cuts near the edge of the damaged square *(right, above)*, taking care not to cut into any adjoining squares. Then take a hammer and butt chisel and remove the rest of the damaged square *(right, below)*. Chip out as much of the remaining adhesive as possible, so that the replacement block will sit flush with the surrounding floor using a butt chisel or cold chisel. You can also try using an adhesive-removal product recommended by your dealer.

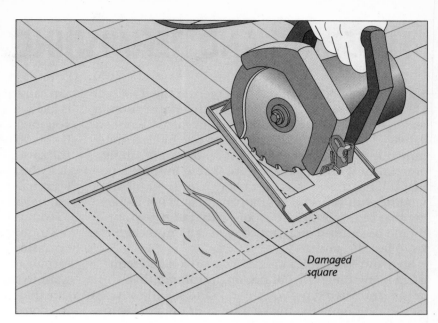

Damaged square

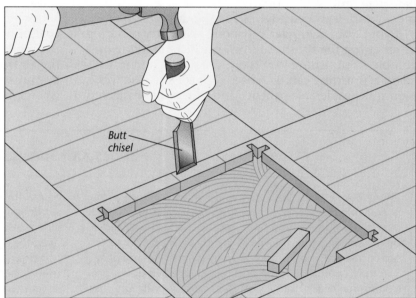

Butt chisel

2 ◀ Installing the new square

If you're installing tongue-and-groove parquet, remove the bottom lip of the grooved edge with a hammer and chisel or table saw. Lightly sand the edges of the square to ensure a good fit. Spread adhesive on the subfloor or back of the tile with a notched trowel and set the new piece in place. Tap it lightly with a mallet without marring the surface. If you get adhesive on adjoining squares, quickly wipe it off with a damp cloth.

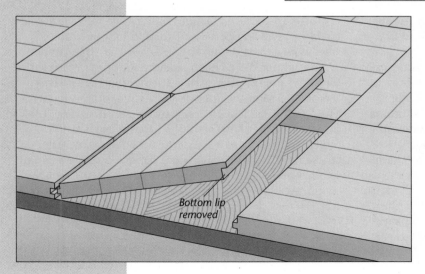

Bottom lip removed

FINISHING AND REFINISHING

New wood floors are sanded to prepare them for sealing, staining, and finishing, but the most common reason for refinishing is simply to rejuvenate a worn and damaged floor, or to give new life to one hidden under carpeting for many years. The procedure for finishing or refinishing is the same, except that refinishing a floor that still has much of its finish may require an initial sanding with very coarse sandpaper that is not required in other cases. CAUTION: Floors with lead-base paint should be stripped, not sanded; you can buy a chemical kit to test the paint.

Standard ¾-inch flooring can be sanded and refinished many times. If your flooring is thinner than this or is a laminated type, you should sand as little as possible. To determine the thickness of your floor, take up a baseboard and shoe molding or a heating grate and measure the exposed edge of the flooring.

If your floor is in relatively good shape, try patching or renewing the finish *(page 58)* before embarking on a complete refinishing job. The specific equipment needed for refinishing can be rented *(below)*. Still, the work requires considerable patience and care; a single misstep can cause irreparable damage. And a refinishing project is likely to be time-consuming, messy, and disruptive— you might want to have the job done by a professional. If you do decide to pay to have the job done, you can still save money by completing the preparatory work described opposite yourself. Then, get a couple of professional estimates for sanding and finishing only. The estimates should cover all materials and labor necessary to complete the job.

Sanding a floor is a very dusty and noisy job. Anyone operating a power sander should wear a dust mask. Use earplugs or earmuffs to protect your hearing. You'll also want to wear soft-sole, nonmarring shoes—and make sure that no one walks on your newly sanded floor. With the pores of the wood exposed, an unprotected floor will soil easily.

In scheduling your project, remember that the first coat of the new finish should be applied the same day the sanding is completed. This will prevent moisture in the air from raising the grain in the raw wood, and dirt from marring the exposed floor. For information on finishes and how to apply them, turn to page 58.

TOOLS AND SUPPLIES

The heavy-duty equipment you'll need to refinish a floor is available at most equipment rental companies. Check over all equipment you rent, and ask that it be tested before hauling it home. You'll need sandpaper in a number of grits, to fit onto both a drum sander and an edger. You may need a hammer and nailset to drive down protruding nailheads, and a putty knife and wood putty to fill holes, dents, and gouges. Choose putty that will blend with the finish or that will take a stain. For applying the finish, you'll need a long-handled lamb's wool applicator, or a paintbrush.

Floor sander: You'll need a basic drum sander, designed specifically for stripping old finishes off wood floors and producing a sanded surface ready for sealing and finishing. Drum sanders come in several models; look for one that has a tilt-up lever that makes it possible to raise the drum off the floor without having to lift the machine. Be sure to get a machine with a dust bag. This won't eliminate dust, but it will cut down the quantity. Check to see that the machine operates on 110 volts and doesn't require a 220-volt outlet. Most sanders have three-prong plugs, so if you don't have outlets that take three-prong plugs, you'll need an adapter. CAUTION: If you have to use an adapter, make sure it's grounded properly.

Ask the rental agent to show you how to install the sandpaper and make sure the necessary wrenches are provided; also ask for tips on operating the machine—*if not operated correctly, the sander can damage your floor.*

Edger and hand scraper: An edger is a disc sander used to sand areas that can't be reached with a drum sander (next to walls, for example). Use a hand scraper for cleaning out corners and other tight areas.

Buffer: A professional floor buffer—the type with a single large revolving pad—is needed to buff the floor with steel wool after the stain and/or sealer have been applied.

Sandpaper: You'll need a different grit of sandpaper for each sanding—the smaller the number, the coarser the grit. To sand the floor of an average room—about 14 by 16 feet—you'll need about two sheets of each grit for the drum sander and four edger discs of each grit. Worn paper won't work as a substitute for a finer grit. Very coarse (20 to 30 grit) open-coat paper may be needed to remove some old finishes or where a floor is particularly rough—choose 20 for paint. Then continue with a coarse grit. For new floors, all parquet floors, and old floors with most of the finish worn off, start with a coarse grit: 36 to 40. A medium grit (50 to 60) is needed for the second sanding, and fine grit (80 or 100) for the final sanding.

Preparing the floor

TOOLKIT
- Hammer
- Butt chisel
- Nailset

1 ▶ Removing the trim

Empty the room of everything, including draperies. Remove all floor grates, doorstops, and any other fixtures. To contain the dust, seal doors to other rooms with masking tape, where possible, and tack plastic inside heating registers.

Take up the shoe molding along the baseboards with a hammer and butt chisel. Use a wood block to protect the baseboard—you may need to use a second chisel between the shoe molding and the floor *(right)*. Number each piece of shoe molding with chalk or a pencil for easy replacement.

2 Inspecting and repairing the floor

Before you refinish a floor, all problems should be corrected. Loose boards should be nailed down, and damaged boards or squares replaced *(page 48)*. Your floor should be inspected for structural problems *(page 21)*, and squeaks should be silenced *(page 22)*. Set any protruding nails with a nailset. Finally, sweep or vacuum the floor thoroughly.

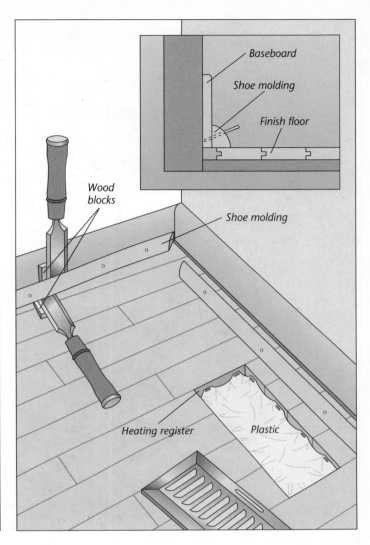

Baseboard
Shoe molding
Finish floor
Wood blocks
Shoe molding
Heating register
Plastic

Sanding a strip or plank floor

TOOLKIT
- Drum sander and its wrench or open-end wrench (optional)
- Edger and its wrench
- Nailset
- Putty knife
- Hand scraper
- File
- Sanding block

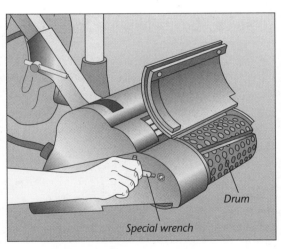

Special wrench
Drum

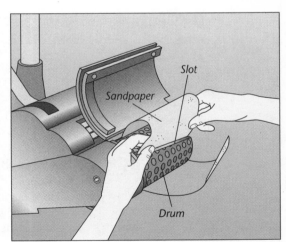

Slot
Sandpaper
Drum

1 Setting up the drum sander

Making sure that the drum sander is unplugged, load the drum with the coarse-grit sandpaper. Use the special wrench provided with the sander to open the slot in the drum *(above, left)*. Wrap the sandpaper around the drum, tucking the ends into the slot *(above, right)* so that the paper is tight. On some sanders, there may be a device to crease the sandpaper. To close the slot, turn the wrench in the opposite direction.

Some sanders have a clamp nut at each end of the drum that must be loosened and tightened with an open-end wrench.

2/3 | **1/3**

Unsanded area

Sanded area

Drum sander

2 Doing the first sanding

For this pass, use a coarse-grit sandpaper. If the pass doesn't remove almost all the old finish, try a very coarse grit and then make another pass with the coarse grit. Take off the minimum amount of wood while removing all of the old finish. In general, sand with the grain. However, if you're faced with an exceptionally rough floor or a floor with cupped boards, make a first pass on the diagonal and then repeat with the grain.

When you're ready to make your first pass, lower the drum slowly and move out the instant it touches the floor. If you allow the drum to bite into the floor before beginning to move, you'll gouge the floor. The forward rotation of the drum will pull the machine forward. Keep it in hand and move forward at a steady pace *(left)*. As you approach the end of your first run, be ready to lift the drum off the floor while the sander is still moving forward. If you try to stop the sander and then lift the drum, you won't be able to act quickly enough to avoid gouging the floor.

When you've completed one pass, shift the sander over to begin the next pass, with the drum overlapping the last cut by 2" to 3". On the return trip, pull the sander backwards at the same steady rate. Continue until you've made your way across the floor. Sand two thirds the width of the room, then start from the other side of the room and work toward the center to sand the remaining third, as shown in the inset.

3 Sanding around the borders

When you've finished sanding as much of the floor as you can with the drum sander, it's time to load the disc sander with the same coarse-grit sandpaper and sand the areas the drum sander was unable to reach. Work on these areas until the disc has removed about the same amount of material as was removed by the drum sander. When you've completed the first sanding, vacuum the floor.

Edger

4 Doing the second sanding

Load the drum sander with a medium-grit sandpaper and make a second pass over the whole floor. Then, with medium-grit paper on the edger, go over the edges once again. Set any exposed nailheads using a nailset and fill these and any other holes with wood putty.

Floor scraper

5 Scraping the edges

Before the final sanding, clean out any tight spots—such as corners and areas around pipes and radiators—that you were unable to reach with the drum sander or edger. Use a hand scraper, pulling toward you while exerting downward pressure (left). If possible, scrape with the grain. Keep a file handy—the scraper should be sharpened frequently to make it an effective tool. When you've scraped away the old finish, use a sanding block with a medium-grit sandpaper to go over the scraped area. Finish off with a fine sandpaper of the same grit you plan to use for the final floor sanding.

6 Doing the final sanding

Using fine-grit sandpaper, go over the entire floor one last time with the drum sander and edger. When the final sanding has been completed, clean up the dust with a vacuum cleaner. Make a final sweep of the floor with a rag moistened with mineral spirits; if you're applying a water-based finish, use a dry cloth.

Check the surface for any blemishes that may have escaped notice earlier, and take care of them now. If any nail holes are showing, fill them with wood putty matching the final finish color, and sand lightly with a fine-grit sandpaper.

CAUTION: In disposing of the debris from the dust bag of a drum sander, do not dump the dust directly into a closed garbage can. The dust will be warm and will contain wax and varnish residue—it could catch fire. Put it outside to cool and then transfer it to a closed container.

Sanding a parquet floor

Sanding on the diagonal

The basic directions for sanding a strip or plank floor (page 55) apply also to a parquet floor. The only difference is that with parquet, all three sandings should be done in a diagonal pattern. This compensates for the fact that the grain in these floors runs in several directions. Also, a parquet floor should never be sanded with a very coarse-grit sandpaper.

Make the first pass with the drum sander from one corner diagonally across the room at a 45° angle. The second sanding—with medium-grit sandpaper—runs across the other diagonal of the room to produce a crisscross pattern, as shown below. The final, fine sanding follows the direction of the first sanding.

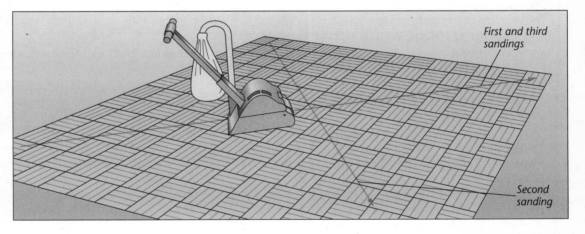

First and third sandings

Second sanding

FINISH APPLICATION

When you have a perfectly clean and exposed floor, you can go ahead and begin the process of applying the final floor finish. Remember that the first coat should be applied the same day the final sanding is finished. This will keep the exposed wood from getting dirty and absorbing any moisture that may raise the grain.

Study the information given below to choose the type of finish you want to apply. Regardless of what products you've selected, read through the manufacturer's instructions carefully before applying any finish—they may vary from those given opposite. It's wise to test the products you've chosen in a closet or inconspicuous area before applying them to the entire floor.

Many prefinished parquet squares are made with rounded or beveled edges, making it impossible to remove the old finish between tiles by machine sanding. If you're refinishing this type, you'll want to choose a finish that will match or blend with the old finish remaining on the tile edges—consult your dealer.

Allow the floor sufficient time to dry—plan on about 72 hours—after finishing; then replace the shoe molding, grates, and other fixtures that were removed. If you've used a penetrating sealer, the floor will require waxing. Select a wax made for floors, and follow the manufacturer's instructions for applying it. For more information on caring for your finished floor, see page 47.

FINDING THE RIGHT FINISH

Floor finishes include stains, penetrating sealers, and surface finishes. If you wish to use stain, it is applied first; then apply a penetrating sealer or a surface finish. Products are available that combine the stain and sealer. Some surface finishes require sealing first.

Stain: If you want a color other than natural wood, the first step is applying stain. Stain without sealer needs 8 hours or less to dry, while one combined with a sealer may require up to 48 hours. If you're going to refinish a softwood floor, it may be necessary to seal it before staining, or to use a sealer mixed with stain, to achieve a uniform color. Ask your dealer what products to use and how to use them.

Penetrating sealer: These products actually penetrate the pores and become an integral part of the wood. The finish wears only as the wood wears. The floor can be re-touched in heavy-traffic areas without creating a patched appearance *(below)*.

Penetrating sealers can be used either clear or tinted, or can be purchased in premixed stain colors. Normally, you'll need to apply two coats. If you buff a penetrating sealer while it's still wet, it will result in a satin sheen. A final coat of paste or liquid wax—but not a water-base wax—is recommended.

Surface finishes: These provide a clear coating over stained or sealed wood. In general, polyurethanes have replaced the more traditional finishes, such as varnish, shellac, and lacquer. Polyurethane and water-base polyurethane finishes are blends of synthetic resins, plastic compounds, and other film-forming ingredients that produce an extremely durable, moisture-resistant surface. They are a good choice for areas subject to splashed or spilled water. High-gloss or satin polyurethane finishes are available. Polyurethane finishes will amber slightly with age; water-base polyurethanes resist ambering. Other surface finishes are available, but they are harder to apply.

 ASK A PRO

CAN I AVOID REFINISHING MY ENTIRE FLOOR?

Often, finish can be patched or renewed without sanding and refinishing. You'll need to determine what kind of finish you have—penetrating or surface (above). Scratch the surface of an inconspicuous spot with a coin; if it flakes, you have a surface finish.

A penetrating finish is the easiest type to repair. You can use a commercial reconditioning product to strip old wax from a small area or the entire floor. This product also cleans and reseals the wood. Then, you can apply a new coat of wax. With a penetrating finish, the patched areas will blend in well.

For a surface finish, first remove any wax with a wax-removal product. If there is a big buildup of wax, this may be impossible. To check if all the wax has been removed, apply a new coat of finish to an inconspicuous area. After a couple of days, scratch it with your fingernail. If the finish flakes easily, it has not bonded well; you must then sand and refinish the entire floor. If there is no wax, or if you're able to remove it, you can patch a few boards, or renew the entire floor. Dull the surface with sandpaper or steel wool and apply a new coat of finish. With a surface finish, a patched area may show somewhat.

TOOLKIT

To apply a stain:
- Paintbrush
- Floor buffer (optional)

To apply a penetrating sealer:
- Paintbrush or lamb's wool applicator
- Floor buffer (optional)

To apply a surface finish:
- Paintbrush
- Lamb's wool applicator
- Foam painting pad for water-base polyurethane
- Floor buffer

Applying stain

Pour all the stain you'll be using into a bucket and mix it thoroughly. Dip a clean, dry rag (of lint-free fabric) into the stain and spread it liberally over the floor. Near the walls, apply the stain with a clean paintbrush.

Wait 5 or 10 minutes to allow the stain to penetrate the pores of the wood; the length of time the stain is left on before wiping affects the color. Use clean rags to wipe up any excess. Let the floor dry overnight. After staining, some manufacturers recommend buffing. When the stain has dried, buff with a floor buffer equipped with a buffing pad or No. 2 steel wool. NOTE: Never use steel wool if you're going to apply a water-base finish. Follow this with a thorough vacuuming. Now the floor is ready for a coat of either penetrating sealer or surface finish.

Floor buffer

Applying a penetrating sealer

In general, a penetrating sealer is a free-flowing liquid that can be applied with a clean rag, brush, or long-handled lamb's wool applicator. The manufacturers of specific products may have specific recommendations.

Apply sealer liberally, letting it flow into the pores of the wood. Start in a corner or next to a wall to avoid walking over wet sealer. After the sealer has had ample time to penetrate (check the manufacturer's instructions), wipe up any excess with dry rags and let the sealer dry for the recommended length of time. Drying time may be affected by humidity and temperature.

If you're not applying a surface finish, apply a second coat of penetrating sealer. (Some manufacturers recommend buffing after the second coat.) Finally, apply a coat of liquid non-water-based wax.

Applying a surface finish

To apply polyurethane, use a clean brush to put down a coat of finish along the walls and around obstacles. Then use a long-handled lamb's wool applicator to apply the finish evenly over the rest of the floor. With a water-base polyurethane finish, use a foam painting pad. Apply the finish across the grain, and then smooth the brush strokes in the direction of the grain.

Typically, two or three coats of polyurethane are required. Allow the first coat to dry (following label directions); then use a floor buffer equipped with No. 2 steel wool or a buffing pad to smooth the surface. (Don't use steel wool if you're applying a water-base finish.) Corners and hard-to-reach areas should be smoothed by hand. Clean the floor thoroughly with a vacuum cleaner. Then go over it with a push broom wrapped with a slightly damp towel to pick up fragments of steel wool and dust. Work the broom back and forth across the floor without picking it up. Then clean up the debris that has accumulated in front of the broom. Be thorough—anything remaining on the surface will be sealed in when you apply the second coat of polyurethane. When the floor is clean and dry, apply a second and third coat of finish, working across the grain whenever possible.

RESILIENT FLOORING

For the homeowner who hasn't bought new flooring for several years, looking at what's available today in resilient materials will provide many pleasant surprises. Flooring manufacturers have made major strides in developing new products that are attractive, long-wearing, and easy to install.

This chapter includes a discussion of the two basic types of resilient flooring available—sheet and tile. We'll show you how to choose the right type for your needs *(page 61)*, and tell you which tools and supplies you'll need for installation *(page 62)*. Step-by-step instructions for preparing the surface and installing resilient flooring begin on page 63. You'll find tips on caring for resilient flooring on page 69, and information on repairs beginning on page 72.

For a more finished look, sheet flooring should fit snugly around doorways and other obstacles. As shown above, it can be slipped neatly under a door casing, which is first trimmed to make room.

CHOOSING RESILIENT FLOORING

Resilient flooring comes either in sheets up to 12 feet wide or in standard 12-inch-square tiles. Sheet flooring can be fully adhered (laid in adhesive over the entire floor), peripherally fastened (attached with adhesive, double-sided tape, or staples only at the perimeter), or laid loosely, like a rug. Although sheet flooring is available in widths up to 12 feet, you may still have to make seams if you're covering a large area.

Individual tiles either come with a self-stick backing or are laid in adhesive. Use a water-base adhesive, and water to wipe up smears or excess adhesive squeezed up between tiles; the older, solvent-base adhesives are being taken off the market because of their toxic fumes.

The properties of various types of resilient flooring are discussed in detail on page 9. Both sheets and tiles are available with smooth or textured surfaces, in plain colors or in patterns. The hundreds of patterns you can buy include authentic-looking imitations of all types of flooring—brick, slate, wood, marble, terrazzo, flagstone, and ceramic tile. Because there are so many choices, visit your local flooring materials dealer and take the time to explore all the possibilities carefully.

Ease of installation should be a major consideration in your choice of flooring. Most manufacturers will provide specific instructions for installing their products and will specify the kind of adhesive to use. Look over these instructions before making a final choice.

Sheet flooring: Resilient sheet flooring can normally be installed with few or no seams—its greatest advantage over individual tiles. This makes sheet flooring particularly practical for bathrooms, kitchens, laundry rooms, and entryways—rooms where water can be spilled or tracked onto the floor.

Peripherally fastened sheet flooring can be installed over surfaces which may not be suitable for fully adhered sheet flooring. Since it is fastened only at the perimeter, it can contract, bridging many irregularities in the underlayment that would be more evident if fully adhered sheet flooring were used.

Working with a large single sheet of flooring is more difficult than working with individual tiles. But modern resilient sheet flooring is much easier to handle than the stiff, brittle floor coverings that were once common.

All resilient sheet flooring is installed in much the same way. The difference is in the way it's attached—it can be completely adhered, attached at the edges only, or not attached at all.

When ordering sheet flooring, you'll need to tell your dealer, at least roughly, the dimensions of the floor you plan to cover. But if you provide a fairly detailed diagram of the room, the dealer may be able to give you a head-start on your installation by making the first rough cuts on your sheet flooring.

If you're planning to install sheet flooring in a large room, you may have to join two sheets together by making a seam. Your dealer will be able to help you decide how much flooring you'll need, how best to arrange adjoining sheets to cover the floor with the minimum amount of material, and how to ensure that the pattern in adjoining sheets will match. To install sheets with seams, refer to page 66 so you'll know what's involved before ordering.

You should plan to have your sheet flooring delivered at least two or three days before you will be installing it. Sheet flooring will expand and contract as the temperature rises and falls, so new flooring should be stored in a warm, dry place—ideally, in the room where it's to be installed.

Tile flooring: Modern resilient tile (made of solid vinyl, vinyl composition, polyurethane, cork, or rubber) has made once-popular asphalt tile obsolete. Builders and homeowners have turned to newer materials that are easier to maintain and last longer. Rubber tile is not normally used in homes; it usually comes in a more limited range of styles than other tile. But, if you need extra durability, you might want to investigate this option.

Tiles are traditionally set in adhesive, but they can be bought with self-stick backing. This adds slightly to the cost, but for many homeowners, the ease of installation is worth the extra investment.

Resilient tiles are generally available in a standard 12-inch-square size. Once you've chosen the tile you want, your dealer will be able to tell you exactly how many tiles you'll need to cover your floor, based on the room dimensions that you provide.

When you order, ask your flooring materials dealer what the policy is for returning unused tiles. Unless you've ordered tile not normally carried by your supplier, full cartons can usually be returned for credit. Some dealers will accept returns of odd amounts if the tile is one of their popular styles.

 ASK A PRO

HOW MUCH RESILIENT TILE WILL I NEED?

When ordering, be sure to allow for spoilage. If you're planning to cover a large floor (over 1000 square feet) you should leave a spoilage margin of about 3%. If you're covering a small floor that's between 50 and 100 square feet, you'll want to order 10% to 12% extra. In any case, it's a good idea to put aside a few pieces in case it's ever necessary to replace damaged tiles.

TOOLS

For installing resilient sheet or tile flooring, you'll need to use many of the standard tools that are illustrated below. You can rent a floor roller from your flooring materials dealer or from a tool rental company.

Always follow basic safety guidelines: wear work gloves when handling sharp or rough materials; wear eye protection when hammering or striking; wear rubber gloves when handling harmful products. For comfort, you may want to wear knee pads when kneeling.

Be sure to follow all safety and operating instructions when using power tools; carefully read the owner's manual provided with the tool.

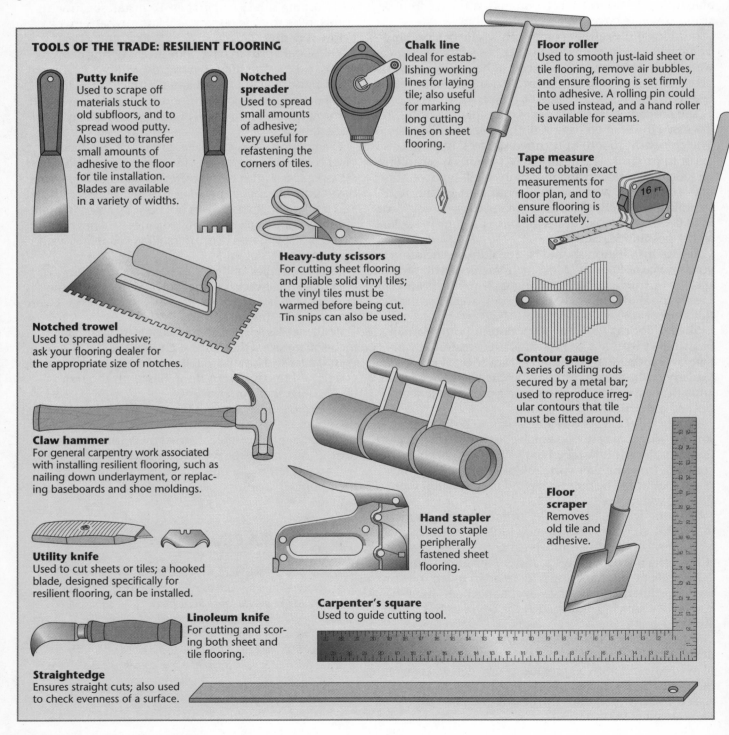

TOOLS OF THE TRADE: RESILIENT FLOORING

Putty knife
Used to scrape off materials stuck to old subfloors, and to spread wood putty. Also used to transfer small amounts of adhesive to the floor for tile installation. Blades are available in a variety of widths.

Notched spreader
Used to spread small amounts of adhesive; very useful for refastening the corners of tiles.

Chalk line
Ideal for establishing working lines for laying tile; also useful for marking long cutting lines on sheet flooring.

Floor roller
Used to smooth just-laid sheet or tile flooring, remove air bubbles, and ensure flooring is set firmly into adhesive. A rolling pin could be used instead, and a hand roller is available for seams.

Tape measure
Used to obtain exact measurements for floor plan, and to ensure flooring is laid accurately.

Heavy-duty scissors
For cutting sheet flooring and pliable solid vinyl tiles; the vinyl tiles must be warmed before being cut. Tin snips can also be used.

Notched trowel
Used to spread adhesive; ask your flooring dealer for the appropriate size of notches.

Contour gauge
A series of sliding rods secured by a metal bar; used to reproduce irregular contours that tile must be fitted around.

Claw hammer
For general carpentry work associated with installing resilient flooring, such as nailing down underlayment, or replacing baseboards and shoe moldings.

Hand stapler
Used to staple peripherally fastened sheet flooring.

Floor scraper
Removes old tile and adhesive.

Utility knife
Used to cut sheets or tiles; a hooked blade, designed specifically for resilient flooring, can be installed.

Linoleum knife
For cutting and scoring both sheet and tile flooring.

Carpenter's square
Used to guide cutting tool.

Straightedge
Ensures straight cuts; also used to check evenness of a surface.

SURFACE PREPARATION

Resilient flooring can be laid on concrete slabs on grade (directly on the ground) or below grade (underground, as in a basement), on wood subfloors made of plywood panels or individual boards, or directly on an old floor. But because most resilient flooring will conform to irregularities in the subfloor, the finished floor will only be as good as what it's laid over, so it's essential that the subfloor is properly prepared to guarantee a smooth surface. It's also important to protect the subfloor, especially if it's concrete, against moisture.

Before you embark on your flooring project, check the basic floor and supporting structure to make sure it's in good condition. Read the section on checking for structural problems on page 21, and follow the procedures recommended. It's important both to discover and repair any structural problems before you start to install your new floor covering.

If the basic floor structure is in good shape, you can go ahead with your flooring project confident that the new floor will have a long life.

CONCRETE SLABS

If you're laying resilient flooring over a concrete slab (on or below grade), be sure the slab is smooth, level, and clean. Check that it is completely dry and take steps to ensure it stays dry. Any moisture coming through the concrete will eventually cause the flooring to loosen.

Most newly cast concrete floors need at least 90 days to dry. Sometimes, this drying can be accelerated with proper heating and ventilation. Concrete may retain some moisture for as long as two years, so even if you're dealing with an existing concrete floor, test for moisture *(page 32)* before proceeding with your flooring project.

Once you're satisfied that the concrete slab is completely dry, check it for grease, oil, and other foreign material. To remove grease and oil spots, use a chemical garage floor cleaner commonly sold at auto supply stores. Use a putty knife or other flat-bladed tool to chip away any lumps of plaster or other hardened materials you may find on the surface of the slab.

It's important to remove old paint or sealer before laying resilient flooring. You can sand the floor with a drum sander *(page 54)*. Vacuum or sweep as you go.

Use a straightedge to find any low spots in the surface. Fill these uneven areas and any cracks or joints with a latex underlayment compound, or a cement-base leveling compound, following the manufacturer's instructions. If the floor is not level, use one of these compounds over the whole floor to level it. When the compound is dry, brush the patched areas with a dry, stiff bristle brush or wire brush and sweep or vacuum up all loose material.

Finally, cover the surface with a sealer or other moisture barrier; your flooring dealer can suggest the best one for your situation.

EXISTING FLOORS

You can put new resilient flooring over an old floor that's been properly prepared to provide a clean, level base.

Old resilient flooring: Resilient flooring can often be installed directly over old resilient flooring, but only if the old floor is completely smooth, solid (not cushioned), and firmly secured. Check with the manufacturer or your flooring dealer for any restrictions that may apply to the flooring you've selected. For instance, solid vinyl tile should not be laid directly over existing resilient flooring; install plywood underlayment first.

If there is no moisture damage and if surface damage is limited to a few loose tiles or small areas where sheet flooring has worked loose, then repairing *(page 72)* and covering the old flooring may be the answer.

Before installing new flooring over old, clean the surface of the old floor thoroughly, removing any wax or finish. CAUTION: Do not sand the old floor covering as a shortcut to cleaning, as it may contain asbestos fibers; see page 72 for information on the dangers of asbestos.

Old resilient tile

Floor scraper

You may also need to take up old flooring to make way for the new floor. Keep in mind that if your flooring was laid before 1986, it may contain asbestos.

To remove resilient sheet flooring, cut it into 6-inch-wide strips with a utility knife. To remove resilient tiles, use a floor scraper as shown on page 63. Be careful not to damage the floor underneath. You must remove all the felt backing, grout, or other material that remains after you've removed the flooring. Most of this can be scraped loose with a floor scraper, or a wide-bladed putty knife, but you may need a solvent to remove stubborn bits of felt. Soften old adhesive by heating it with an iron (to prevent the adhesive from sticking to the iron, cover the area with a piece of paper first).

If you find that the old covering is very difficult to remove, consider covering the floor with new plywood underlayment 1/4 inch to 1 inch thick; see below. In selecting new underlayment, take into account the level of adjoining floors.

Old wood flooring: Most types of resilient flooring can't be laid directly over an existing wood floor because the joints will show through; install plywood underlayment as described below. However, if you're installing peripherally fastened sheet flooring, and the wood floor is level and in good condition, you can install your flooring directly over it.

Other types of old flooring: Resilient flooring should never be installed directly over old ceramic tile, slate, or masonry flooring with an uneven surface. These kinds of old flooring should be removed, if possible, unless they're reasonably level. If the old floor is level, it can be covered with a latex underlayment compound or a cement-base leveling compound. Solid masonry-type floors can be covered with a new concrete slab.

WOOD SUBFLOORS

Wood subfloors are made from plywood panels or individual boards. With either, be sure to clean the surface carefully before applying any floor sealer.

To avoid the damaging effects of moisture, the wood subfloor must be suspended at the correct height (commonly a minimum of 18 inches above the ground) and adequately cross-ventilated underneath.

Plywood subfloors: Preparing a plywood subfloor for new resilient flooring is relatively simple. If the subfloor is new, make sure the panels are securely attached with ring-shank or cement-coated nails, and that the nailheads are flush with the surface. If you've taken up old flooring, make sure to remove all foreign material from the subfloor.

Use wood putty to fill minor indentations and gaps larger than the thickness of a dime between panels; allow the putty to dry thoroughly before sanding it smooth.

Wood board subfloors: Because it is extremely difficult (if not impossible) to make most wood board subfloors sufficiently smooth and level to install a resilient floor covering, they should be covered over with underlayment-grade plywood.

Before you begin, check the subfloor and make any necessary repairs. Renail any loose boards to make the subfloor as level as possible; if boards are badly cupped or bowed, you might have to replace them. Boards that aren't so misshapen can often be sanded smooth with a drum sander *(page 54)*. Minor indentations in the subfloor need not be repaired because the plywood underlayment will cover them.

Underlayment: Use 4x4-foot or 4x8-foot sheets of underlayment-grade plywood. As you install the panels, keep these three points in mind:
• Stagger joints between panels to avoid having four corners meet *(page 33)*.
• Leave gaps about the thickness of a dime between panels to allow for expansion of the material.
• Allow 1/8-inch clearance between panels and walls or baseboards.

Fasten the panels with 1 1/4-inch ring-shank nails or 1 1/2-inch cement-coated nails spaced 3 inches apart along the edges and 6 inches apart across the face of each panel. Always start nailing in the center and work out toward the edges.

 ASK A PRO

HOW DO I PREPARE THE ROOM FOR NEW FLOORING?

Remove all furnishings from the room, and take up any heat register or vent covers. If the baseboards are faced with shoe molding, only these need to be removed, unless you're re-placing the baseboards with cove base (page 67). Baseboards should be removed only if they can be loosened without damaging the walls or doorjambs. Label or number the pieces of shoe molding, for easier replacement.

You should also decide now between the two options for trimming your new flooring around the room's door casings. The easiest method is to cut away just enough of the bottom of the door casings to slip the flooring underneath (page 69). The other option is to trim the flooring itself to fit flush with the casings. If you want to go the first route, trim the bottoms of the door casings now.

INSTALLING RESILIENT SHEET FLOORING

Installing resilient sheet flooring is as much a matter of deliberate planning and careful trim work as it is of hard labor. Whether you're installing fully adhered, peripherally fastened, or loose-laid flooring (not fastened), the procedures that are involved are basically the same. The general techniques are shown here, but be sure to follow any manufacturer's instructions for your floor covering or adhesive, especially regarding the latter's open time. This is the time that is required for the adhesive to dry—you'll have to install your flooring within this time. You may also need to leave the adhesive time to tack up (become sticky) after you spread it but before you lay the flooring.

To ensure a clean, professional-looking installation, draw an accurate floor plan and map out the best approach to installing the flooring, especially if you'll be using more than one sheet. Using this plan, either you or your flooring dealer will make the first rough cuts. Make a scale drawing of the room on graph paper,

using exact measurements. Include the locations of doorways, alcoves, closets, counters, and any other irregularities in the room. Having a partner will help make this job go more quickly. If your room is very irregular, you may want to make a full-size pattern of the floor instead of a scale drawing.

Although resilient sheet flooring can be purchased in widths up to 12 feet, it may be necessary to make a seam between two pieces to cover a large area. Looking at your floor plan, determine how to combine sheets so that you can cover the floor with the minimum amount of material. New seams should not follow seams in existing resilient flooring or joints in the underlayment or subfloor. If the flooring is patterned, you'll need enough to match the pattern at the seams.

For loose-laid flooring, the standard method of seaming two pieces together involves fastening the sheets with double-sided tape, but check what technique and product the flooring manufacturer recommends.

Installing sheets without seams

TOOLKIT
- Tape measure
- Carpenter's square
- Straightedge
- Utility knife or linoleum knife
- Heavy-duty scissors or tin snips

To apply adhesive:
- Notched trowel (for large amounts of adhesive)
- Notched spreader (for small amounts of adhesive)

For stapled flooring:
- Hand stapler

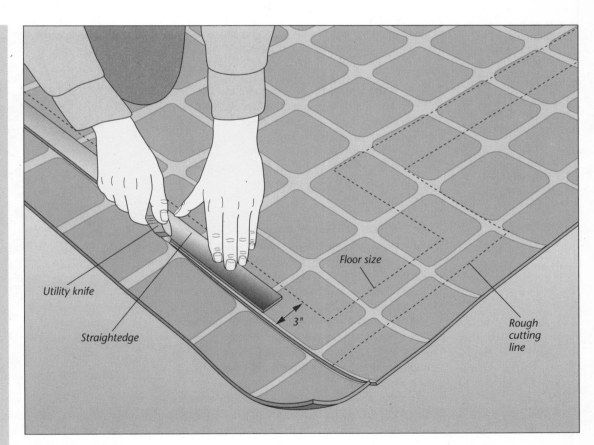

Utility knife

Straightedge

Floor size

3"

Rough cutting line

1 **Cutting the sheet**
Unroll the flooring and transfer the floor plan onto the sheet using chalk or a water-soluble felt-tip pen, a carpenter's square, and a long straightedge. Using a knife and a straightedge for straight cuts *(above)*, heavy-duty scissors or tin snips for curves, cut the flooring 3" oversize on all sides; accuracy is critical.

2 Fastening

For fully adhered flooring, use a notched trowel to apply the adhesive to the floor—over the entire floor, or in stages as the flooring is unrolled. Be aware of the time it takes for the adhesive to dry (open time). Lay the longest edge of the sheet against the longest wall, allowing the 3" excess to curl up that wall and the two adjacent walls. If the entire floor has been covered with adhesive, slowly roll the flooring out across the floor, setting it firmly into the adhesive as you go. If you're spreading adhesive in stages, work your way across the floor, unrolling the flooring as you go. Trim the excess and finish the job *(page 68)*.

For peripherally fastened flooring, lay the sheet in the room, aligning the pattern squarely, and allowing the excess to curl up the walls. Weight the sheet, then trim *(page 68)* and attach it *(step 2, below)*.

If you're installing loose-laid flooring, simply roll out the sheet, aligning the pattern squarely in the room, and shift it into position. Weight the sheet, and trim the edges and corners, as described on page 68.

Installing sheets with seams

TOOLKIT
- Tape measure
- Carpenter's square
- Straightedge
- Utility knife or linoleum knife
- Heavy-duty scissors or tin snips

For adhered flooring:
- Notched trowel (for large amounts of adhesive)
- Notched spreader (for small amounts of adhesive)
- Hand roller or rolling pin

For stapled flooring:
- Hand stapler

1 Marking the sheets

Unroll the flooring in a large, clean area. Using chalk or a water-soluble felt-tip pen, a carpenter's square, and a long straightedge, transfer the floor plan you've made to the flooring material, as shown on page 65. On patterned flooring, be sure to leave the margins necessary to match the pattern on adjoining sheets at the seam. If your flooring has a simulated grout or mortar joint or any other straight line, follow the manufacturer's instructions for cutting it; you can either cut the seam along the midpoint of the printed joint on both sheets, or cut the joint off one sheet and leave it on the other. Ask your flooring materials supplier if you'll need extra material to reverse the sheets at the seam for the design that you've selected.

2 Cutting and installing the sheets

Using a knife, a straightedge, and heavy-duty scissors or tin snips, cut the piece that requires the most intricate fitting roughly 3" oversize on all sides. For peripherally fastened flooring, don't leave excess at the seam line; cut the seam accurately following the manufacturer's instructions.

For fully adhered flooring, spread adhesive for the first sheet with a notched trowel, but stop about 8" from the seam. Position the longest edge of the first sheet against the longest wall, allowing the 3" excess to curl up the wall, and roll the sheet out into the adhesive. Next, cut the second sheet and position it to overlap the first sheet at least 2", *(right)*, then keeping the overlapped seams in place, roll up the sheet and spread the adhesive over the remainder of the floor, stopping a few inches from the edge of the first sheet. Reposition the sheet, aligning the design perfectly, and roll the flooring out, setting it into the adhesive.

For peripherally fastened flooring, position the first sheet in the room, lining up the pattern squarely with the walls. Position the second sheet so that the pattern matches at the seam line. Complete the seam *(step 3)*, and trim, then fasten the edges. Use staples if the underlying surface is wood; make sure they're close enough to the wall to be covered by the molding. Use adhesive if the subfloor is concrete; spread it in a 3" band around walls and around anything the flooring is cut to fit around.

For loose-laid flooring, lay the first sheet in place, then cut the second sheet and position it carefully to overlap the first one by about 2". Next, lift it up carefully and secure it to the floor with two or three pieces of double-sided tape so it won't move while you cut.

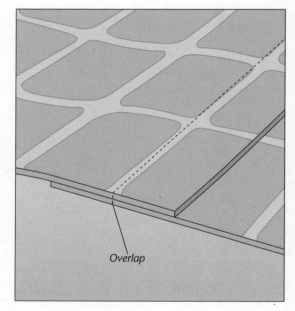

Overlap

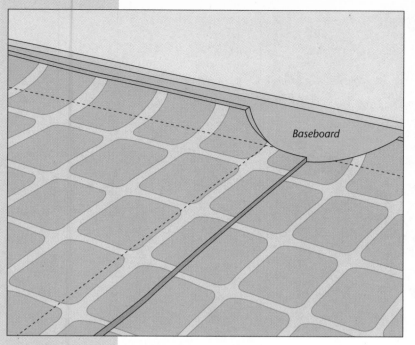

Baseboard

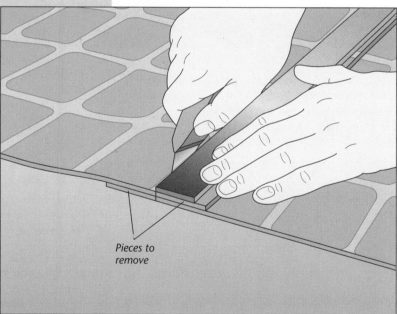

Pieces to remove

3 Making the seam

For fully adhered and loose-laid flooring, trim away excess material at each end of the seam in a half-moon shape, so the ends butt against the wall and the seam lies flat *(left, above)*. Cut off the excess material using a long straightedge and a utility knife. Make a straight cut following the joint line, as recommended, down through both sheets of flooring *(left, below)*.

For fully adhered flooring, roll back the edges of the flooring and lay a strip of double-sided tape on the floor where the seam will be. Spread adhesive on the floor, covering the tape. Peel the backing off the tape and roll the sheets back into position, one side at a time, lining up the pattern. Roll the seam with a hand roller or rolling pin. When the seam is dry and dirt-free, use the seam sealer recommended by the flooring supplier for your specific type of flooring, to fuse the two pieces.

For peripherally attached flooring, with the seam lines cut, roll back the edges of the flooring and spread adhesive on the exposed floor where the seam will lie. Roll the flooring back into position, one side at a time, carefully aligning the pattern. You may need to stretch or compress the material. Use masking tape to hold the seam in position; roll the seam with a hand roller or rolling pin, then wipe up excess adhesive. Trim the sheet at corners, walls, and doorways *(page 68)*. Fasten the perimeter *(step 2)*. When the adhesive is dry, remove the tape, clean any adhesive out of the seam with a utility knife, and apply the recommended seam sealer.

For loose-laid flooring, roll the flooring away from the seam; then, stick double-sided tape to the floor, removing one side of the backing paper as you go; stop 5" from each wall. Peel off the remaining backing paper and press the flooring down firmly, keeping the edges close together.

NOTE: Any adhesive in the seam may prevent the seam sealer from functioning properly.

ASK A PRO

DO I HAVE TO USE BASEBOARDS?

An alternative to baseboards and shoe molding is cove base (also known as wall base). Cove base is flexible and comes in rolls or standard 4-foot lengths. It is installed using the same tools required for resilient flooring and special adhesives made for use on vertical surfaces. Cove base is fastened directly to the wall, with the lower edge resting on the resilient flooring; don't attach it to the surface of the floor. Spread the adhesive with a notched trowel, and allow it to set as directed; starting at an inside corner, press the base firmly in place until it holds. At outside corners, stretch the vinyl around the corner, pressing it firmly against the wall. At inside corners, cut the vinyl to fit, mitering the ends with a utility knife. Or, score the back of the vinyl and bend it to fit snugly to form an inside corner without joints.

Trimming the flooring and finishing the job

TOOLKIT
- Utility knife or linoleum knife
- Straightedge
- Handsaw

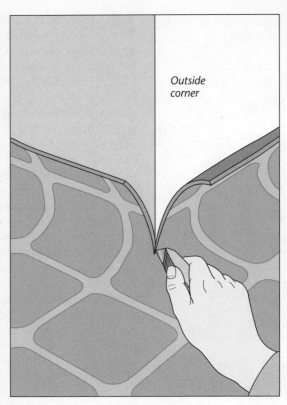

Outside corner

Inside corner

1 Trimming to fit corners

Make a series of relief cuts at all outside and inside corners so the flooring will lie flat on the floor. At outside corners, start at the top of the excess flooring and cut straight down to the point where the wall and floor meet *(above, left)*. At inside corners, cut the excess flooring away with diagonal cuts from top to bottom, gradually trimming it until the flooring lies flat in the corner *(above, right)*.

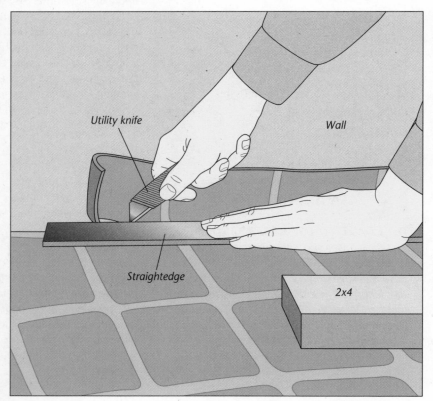

Utility knife

Wall

Straightedge

2x4

2 Trimming at walls

Using an 18" to 24" piece of 2x4, press the flooring into a right angle where the floor and wall join. Lay a straightedge along the wall and trim the flooring with a utility knife; leave a gap of about 1/8" between the edge of the flooring and the wall so the material can expand without buckling. If you're planning to attach cove base, be sure it will overlap the edge of the flooring by at least 1/4".

3 ▶ Trimming around doorways

The flooring should fit snugly around doorways and any obstacles where there will be no molding to provide a finished appearance. Around a doorjamb, the most effective way to hide an exposed edge is to cut away just enough of the door casing *(right, above)* to slide the flooring underneath *(right, below)*. It's best to cut the door casing when you're preparing the room—before laying your flooring. The other option is to trim the flooring so it fits against the door casing, but it's harder to achieve a professional-looking job this way.

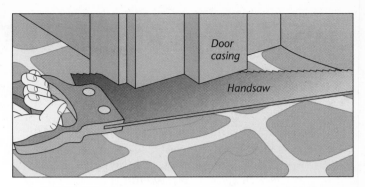

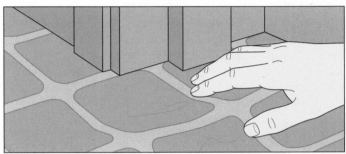

4 ▶ Cleaning up and reinstalling trim

With the flooring in place, clean the surface. The manufacturer may recommend a cleaner that won't harm the flooring. Wipe up any adhesive on the surface; for dried adhesive, first use soap and water, then try mineral spirits. CAUTION: Mineral spirits are highly flammable.

For fully adhered flooring, make sure the sheets are set firmly into the adhesive. Using a floor roller, start at the center of the room and work out any air bubbles that may remain.

When the floor is clean, flat, and well settled, you can replace any baseboards that have been removed. Then reattach the shoe molding, leaving a 1/32" to 1/16" gap between the floor and the bottom of the shoe molding. Always drive nails through the molding into the baseboards, and not down through the flooring; in this way, the flooring can move slightly without buckling. Finally, replace any vent covers or fixtures that were removed.

CARING FOR YOUR RESILIENT FLOORING

The two greatest problems of maintaining resilient flooring are infrequent cleaning and overpolishing. Grit particles have an extremely abrasive effect on the floor's surface; these particles make thousands of small cuts that dull the surface of the floor. You can prevent surface wear and damage by sweeping or vacuuming the floor regularly.

Combined with regular sweeping, an occasional cleaning with a mild cleaning solution, such as clear or sudsy ammonia, is sufficient to keep resilient flooring in good condition. Use a clean sponge mop for washing and rinsing; avoid flooding the surface.

To remove a tough stain, use soap and water on a soft cloth or nylon pad; start at the edges. If water doesn't work, try liquid floor cleaner, rubbing alcohol, diluted liquid chlorine bleach, or mineral spirits. Wipe up the residue with a damp cloth, and dry thoroughly. To remove a sticky residue, first chill it with an ice pack, then scrape it up gently with a plastic spatula. Wipe the area with a damp cloth, and dry.

Many resilient floor coverings have no-wax surfaces, and don't need polish; others may require polishing as part of regular maintenance. Follow the manufacturer's instructions for the flooring you've installed.

In general, one-step clean and shine products are not recommended. These combine a detergent and a polish; the theory is that the detergent loosens dirt particles, which are retained in the applicator as the polish is spread. However, often the applicator can't absorb all the dirt, so it's spread and trapped in the polish. Over time, this adds damaging abrasive materials to the surface.

INSTALLING RESILIENT TILE FLOORING

When the subfloor is properly prepared, you can establish your working lines—the guides you'll use to place your tile accurately. There are two methods of laying tile—from the walls or from the center of the room. If you start from the center, you'll have to cut border tiles around the perimeter of the room; if you start from the walls, you can have full tiles on two walls and cut tiles on the other two. To work from the walls, your room must be square and your walls straight. Turn to page 82 for information about establishing working lines from the walls, and page 43 for establishing working lines from the center of the room. The instructions below are for laying tile from the center of the room.

To make straight cuts in brittle resilient tile, score the tile with a knife; snap the tile at this line. Irregular cuts can be made with scissors or tin snips if the tile is warmed (use a hairdryer). The more pliable types of tiles are always warmed and then cut with scissors or tin snips. Tiles should be warm, but not hot to the touch.

If you're using adhesive, be aware of the adhesive's open time—the time it takes for the adhesive to dry. You may also need to leave the adhesive time to tack up, before you begin placing your tiles; follow the label directions. You'll have to lay your tiles within the open time, so don't spread adhesive over a larger area than you can cover at once.

SEQUENCES FOR PLACING TILES

Two effective patterns for laying tile are illustrated below. The pattern that you choose will depend on the adhesive's open time. Method A is better when you're working with an adhesive with sufficient open time or if you're using self-stick tiles. Method B is a better choice when you have to work quickly.

Method A

			13			
		14	7	12		
	15	6	4	8	11	
16	5	3	1	2	9	10

Method B

	11			
	10	12		
	4	9	13	
	3	5	8	14
1	2	6	7	15

Laying resilient tiles

TOOLKIT
- Notched trowel
- Broad putty knife or taping knife
- Utility knife or linoleum knife
- Heavy-duty scissors or tin snips
- Contour gauge or compass
- Claw hammer

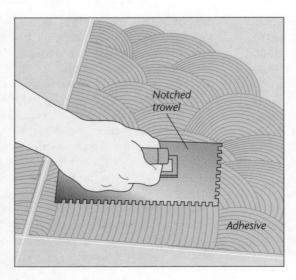

Notched trowel

Adhesive

1 Applying adhesive
If your tiles are self-stick, you're ready to place them. If you're using adhesive, you can scoop out the adhesive with a notched trowel, or transfer a small amount with a broad putty knife or taping knife to the spot on the floor where you plan to lay the first tile. Take the notched trowel, and, using light pressure, spread the adhesive evenly across the floor *(left)*. If you can't see your working lines through the ribbons of adhesive, don't cover the lines. Until you have a feel for how many tiles you can set within the adhesive's open time, use the adhesive sparingly.

NOTE: Don't forget to leave the adhesive time to tack up before laying your tiles. If glue squeezes out between the tiles, you haven't left enough time.

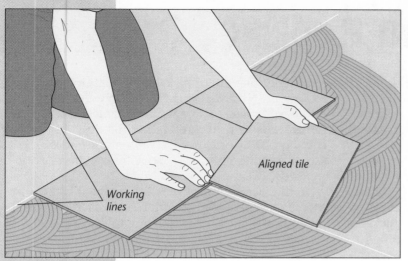

2 Placing the tiles

Carefully line up the first tile at the point where the working lines cross, then follow the sequence you've chosen. For self-stick tiles with arrows on the back, follow the manufacturer's instructions to point the arrows either in the same, or in alternating directions. To place each tile, align an edge with a working line or an adjacent tile, then let it fall into place *(left)*. Never slide a tile into position, or you'll force adhesive up between the tiles. When a tile has been positioned correctly, press it down firmly into the adhesive. Kneel on a piece of plywood so you don't disturb freshly laid tile. When you've finished laying the floor, bed the tiles in the adhesive with a floor roller or rolling pin.

Working lines

Aligned tile

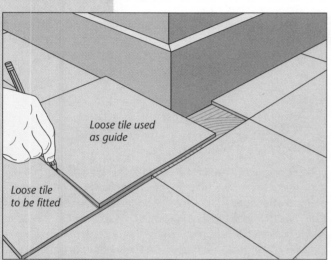

Loose tile used as guide

Loose tile to be fitted

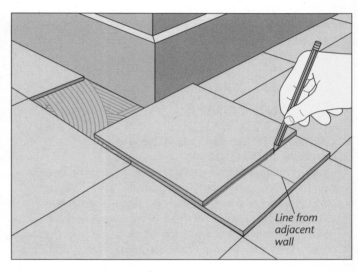

Line from adjacent wall

Piece to be removed

Corner tile

Pencil lines

3 Marking and cutting border tiles

To mark border tiles for straight cuts, place a loose tile squarely on top of the last full tile nearest the wall; then place another tile on top of it, so it butts against the wall. With a pencil, draw a line on the lower tile along the edge of the upper tile *(above, left)*. When the lower tile is cut, it will fit in the border.

Cut L-shaped tiles for outside corners in the same way. For the first line, proceed as above for border cuts. Next, place the tile to be cut on top of the last full tile closest to the adjoining wall. Place a loose tile on top and draw another line *(above, right)*. Cut along the lines to remove the corner piece *(left)*. To fit tile around doorjambs, cut a pattern out of heavy paper or cardboard and trace it onto a tile; or use a contour gauge or a compass to transfer the pattern onto a tile; align the tile with its row first.

4 Finishing the job

Check that all tiles are firmly set and smooth. Use soapy water to clean up any dry adhesive (use mineral spirits for stubborn spots). Replace any baseboards that have been removed and reattach the shoe molding, driving the nails into the baseboards, not into the flooring. To install cove base instead of baseboards and shoe molding, turn to page 67.

REPAIRING RESILIENT FLOORING

Resilient flooring is, in general, easy to repair; however, it's a good idea, before going ahead with any repair project, to determine if the trouble on top of your floor is caused by more serious problems below the floor. Read the section on possible causes of damage (*opposite*), and investigate before undertaking repairs.

Matching replacement materials to your existing floor may be a problem, especially if your flooring is fairly old. Even if you have leftover flooring from the original installation, you may find that it differs in color and thickness from the old flooring, and will look too conspicuous. If the flooring to be replaced is confined to one area, consider replacing the damaged flooring with a contrasting or complementary pattern or color, to create a new design. If you can't make repairs that are visually acceptable, it may be time to replace your old flooring completely.

Newer resilient flooring is likely to last longer than old flooring—it is made to be more durable. You're more likely to have to repair it as a result of damage (cuts, burns, marks made by furniture), rather than because of wear. You'll probably only need to replace a few tiles, or patch a small area, and finding replacement material should be relatively easy.

On the following pages are step-by-step instructions on how to make the most common surface repairs. You'll need some or all of the following tools: a utility knife, a notched trowel or spreader, a putty knife, a straightedge, and an iron or a propane torch fitted with a flame spreader. Use a water-base adhesive; the manufacturer of your flooring may recommend a specific adhesive for repairs. Use books, bricks, or other heavy objects to weight the flooring while the adhesive is setting (usually overnight).

PLAY IT SAFE

CAUTION: THERE MAY BE ASBESTOS IN YOUR OLD FLOORING

Resilient flooring manufactured before 1986 may contain asbestos fibers, which are hazardous to health if released into the atmosphere. Asbestos is no longer used in flooring materials, but it may be in the flooring already in your home. The only way to know for sure whether your flooring contains asbestos is to have it tested at a private lab; therefore, treat any resilient flooring from before 1986 as if it does contain asbestos. DO NOT sand it. If the flooring is in good shape, you may be able to lay your new flooring directly over it; otherwise, cover it with 1/4-inch plywood underlayment. If you want to remove the flooring, check with your state health department or your local Environmental Protection Agency office about the regulations in your area, and for information on disposal. Information on the safe treatment of flooring containing asbestos is available from the Resilient Floor Covering Institute in Rockville, Maryland.

Repairing a surface bubble

TOOLKIT
• Iron (optional)
• Utility knife and putty knife (optional)

Flattening and gluing the bubble

First, place an ice pack on the bubble. If this doesn't work, try heat: either a hot, damp cloth or an iron set on medium heat (protect the flooring from the iron with a piece of aluminum foil). With a utility knife, cut into the bubble in a straight line, going all the way across the bubble.

Force new adhesive under the cut with a putty knife (*right*), trying to keep adhesive off the two edges of the slit. Wipe off any excess adhesive, and clean out the slit with a utility knife. Weight down the area overnight, then apply the seam sealer recommended by a flooring dealer.

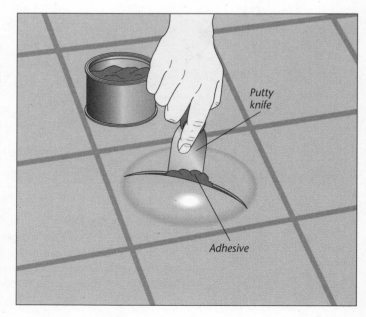

Putty knife

Adhesive

The cause of most surface damage, such as stains or holes, is usually obvious, but sometimes relatively minor surface problems can be traced to more serious underlying causes.

Small bumps in the surface of the flooring may be caused by nails that have worked loose. This sometimes happens because of normal movement in the structure, or because the flooring was originally installed when there was too much moisture in the wood; the nails become loose when the wood dries. If nails have worked loose beneath a small area of the floor, try placing a block of wood over the bumps and tapping lightly with a hammer to drive the nails flush. Be careful not to damage the floor surface. If the nails can't be reseated with light tapping, abandon this approach; you'll have to remove the floor covering to gain direct access to the nails.

If you have a regular pattern of indentations that form Ts, or run for several feet, there may be gaps in the underlying surface; individual boards or plywood panels may separate as the structure settles *(below, left)*. You'll have to take up the flooring and fix the problem.

If resilient tiles have curled at the edges or popped loose, or if sheet flooring has worked loose at the edges or seams, you may have a moisture problem. Moisture in a floor on or below grade may be due to poor drainage outside. Limited moisture problems (loose tiles in one area) may be caused by a plumbing leak *(below, right)*. Fix the problem before you repair the floor.

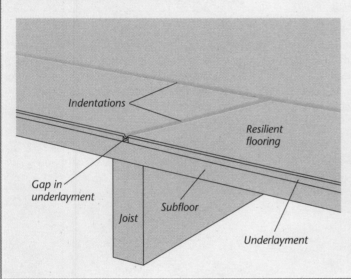

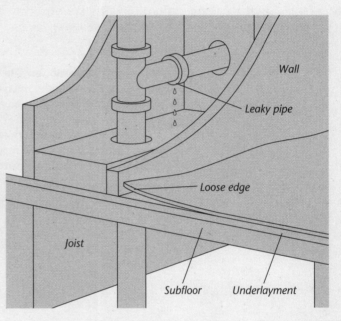

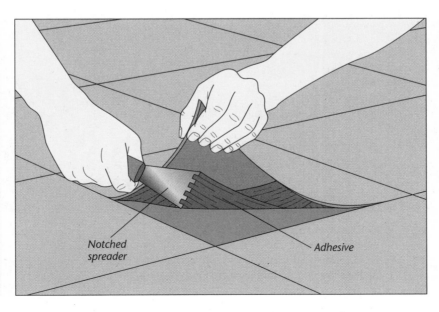

Fastening a curled edge

TOOLKIT
- Iron
- Putty knife
- Notched spreader

Gluing down the edge of a tile
Cover the affected flooring with a piece of aluminum foil, then use a warm iron to soften the adhesive. Scrape off the adhesive with a putty knife. Using a notched spreader, add new adhesive *(left)*.

Press the material firmly back into place, and weight it overnight.

Replacing a damaged resilient tile

TOOLKIT
- Propane torch (with flame spreader) or iron
- Putty knife or old chisel
- Notched trowel

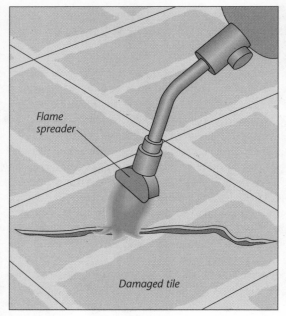

Flame spreader

Damaged tile

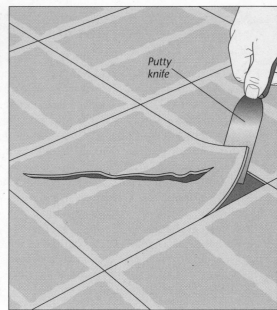

Putty knife

1 **Removing the damaged tile**
Soften the adhesive under the damaged tile using a propane torch with a flame spreader attached *(above, left)* or a warm iron. Pry up the tile with a putty knife *(above, right)* or an old chisel. Be careful not to damage surrounding tiles; it may be better to work from the damaged spot toward the edges.

Remove any remaining adhesive. You may be able to remove individual asbestos tiles by misting the tiles and adhesive with a water from a spray bottle as you remove them; check with your local Environmental Protection Agency or your state health department.

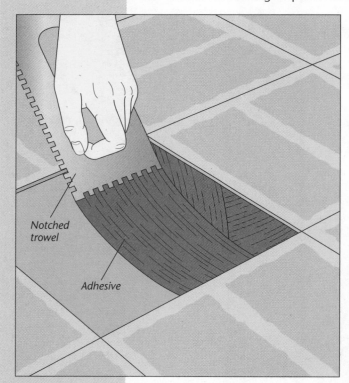

Notched trowel

Adhesive

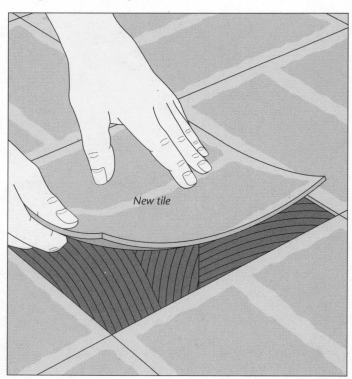

New tile

2 **Installing the replacement tile**
When the floor is dry, apply the adhesive with a notched trowel *(above, left)*. Be careful to keep adhesive off the surrounding tiles. Place the new tile firmly in place *(above, right)*, and wipe up any smeared adhesive. Weight the edges of the tile so it stays in place while the adhesive sets overnight.

Patching resilient sheet flooring

TOOLKIT
• Tape measure
• Utility knife
• Straightedge
• Putty knife
  or old chisel
• Notched trowel

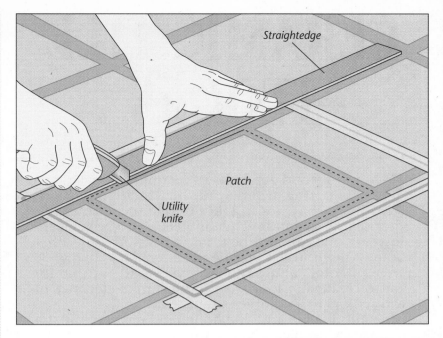

1 Cutting the patch
Measure the damaged area and cut out a replacement piece large enough to cover it. Match the pattern, if there is one, and tape the new piece in place over the damaged area.

Using a utility knife and a straightedge and following the lines in the pattern, cut a patch out of the new piece *(left)*, and down through the old flooring underneath.

2 Removing the damaged piece of flooring
Remove the patch *(right)*, and the unused new material taped to the floor. Use a putty knife or an old chisel to remove the damaged flooring and the old adhesive. Keep the area moist (mist with a spray bottle) if there is any chance that your flooring contains asbestos.

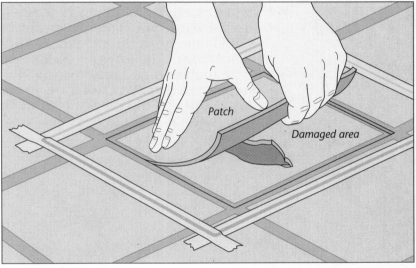

3 Gluing the patch
Using a notched trowel, spread adhesive on the back of the patch that you cut out in step 1 *(left)*. Press the patch firmly in place, and wipe up any adhesive from the surface of the flooring and clean adhesive out of the seam with a utility knife.

Weight the patch overnight, then apply the seam sealer recommended by a flooring dealer.

CERAMIC TILE FLOORS

The beauty and durability of ceramic tile make it an ideal flooring material. Tile has traditionally been used in bathrooms, kitchens, and entryways because it withstands water and is easy to keep clean. However, tile can also be used to dress up any other room of the house. Ceramic tile is available in a wide range of colors and patterns, making it possible for you to create any look—from the simple elegance of earth colors to a complex graphic design.

Installing a ceramic tile floor can be a very satisfying project. In recent years, improvements in grouts and adhesives have made it possible for a careful, patient do-it-yourselfer to create a tile floor of professional quality. In this chapter, we'll help you select the right type of tile for your needs (*opposite*), and choose your materials and tools (*page 79*). Turn to page 80 for step-by-step instructions on preparing the surface, installing the tile (*page 81*), and finishing your project (*page 84*). Finally, we tell you how to replace a broken tile (*page 85*).

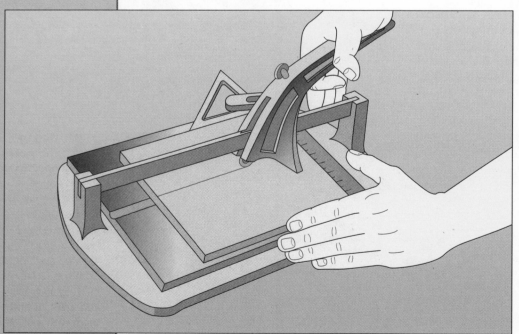

As you lay your ceramic tile, you'll almost always need to cut some tiles to fit along the walls. You can make straight, clean cuts with a snap tile cutter rented from your dealer.

SELECTING CERAMIC TILE

To select the type and pattern of ceramic tile you'll use, visit specialty tile stores or a building supply center with a good flooring materials department. As you shop around, ask to see only those kinds of tile that are manufactured for floor surfaces; wall and counter tiles are thinner than floor tiles and won't hold up under traffic.

You'll find floor tiles available in squares, rectangles, hexagons, octagons, and exotic shapes like ogee and Moorish. Tiles are either glazed or unglazed. Glaze is a hard surface that is applied to the tile before it is fired. The color of the tile is in the glaze; the color of unglazed tiles comes from the clay they're made from, or else from pigment that is mixed with the clay, and is constant throughout. Samples of unglazed tile that appear to have a glossy finish have been treated with a sealer; some are then covered with a glossy floor finish.

Tiles come in a number of types, depending on how they are made. These types include quarry tiles, pavers, glazed tiles, and Mexican tiles.

Quarry tiles: These are made of extruded clay and are available either unglazed in natural clay colors or with a glazed surface. Quarry tiles are manufactured in various shapes and sizes and have a slightly rough surface.

Unglazed quarry tiles should be sealed to make them water-resistant.

Pavers: These tiles are made of clay that is pressed rather than extruded. They are typically unglazed and many need to be sealed to be water-resistant. However, pavers also include porcelain tile, an extremely dense, vitreous tile that does not need to be sealed.

Glazed tiles: Like pavers, glazed tiles are made of pressed clay, but the color is applied to the surface, rather than to the body of the tile. Glazed tile includes mosaics which are available in sheets held together with a plastic or cotton mesh backing. Glazed tiles often have a glossy finish, but some can be ordered in matte or textured finishes that make them less slippery.

Mexican tile: Similar to terra-cotta tile, Mexican, or Saltillo, tile has become popular in the American Southwest and West. Some suppliers classify it as a quarry tile and others as a paver. Relatively soft and unglazed, Mexican tile can be protected with a sealer to keep the surface from powdering.

In addition to these types, consider tiles made of cut stone, such as marble or slate. These are installed in much the same way as ceramic tiles.

When buying tile, keep these tips in mind to save yourself time and money:
• To avoid long delays, select tile that your supplier has in stock or that can be ordered and delivered within a reasonable period of time.
• To make sure you'll have enough tile to complete the job, it's important that you tell your tile dealer the size of the area you wish to cover. Draw a scale map of the floor area on a piece of graph paper, noting all measurements, so your dealer will be able to determine how much tile you'll need and how much it will cost.
• Be generous when estimating and ordering tile. You'll want not only enough tile to complete the job, but also enough extra tile to allow for miscalculations and breakage in cutting trim and border pieces to size. You'll also want to put some tile aside for future use, in case you need to replace a chipped or damaged tile.
• To save money, look for close-out sales. Also, consider making use of "seconds"—slightly blemished tiles; if these are mixed with unblemished tiles, the difference may be unnoticeable.
• When you take delivery of your new tile, check each carton to make sure the colors match.

CARING FOR YOUR CERAMIC TILE FLOOR

Unglazed tile, except for porcelain tile, should be sealed to protect it from moisture and stains. Cement-base grout should also be sealed if it's used on a floor that's going to be subject to a lot of water and staining substances. Consult with your dealer for sealers that are appropriate for your tile and grout. Sealers are either topical or penetrating; topical sealers need to be reapplied periodically, while penetrating sealers protect tile and grout permanently.

For day-to-day care of a sealed tile floor, sweep and then damp-mop with a mild detergent or all-purpose household cleaner. Make sure to sweep away any grit that can be ground into the floor and damage the tile. For tough dirt, you can use a commercial tile cleaner, but some of these cleaners contain harsh acids that can damage the tile and grout—use them sparingly.

If grout becomes stained, scrub it with a toothbrush and household cleaner that does not contain bleach. With colored grout, test the cleaner first in an inconspicuous spot.

To prevent stains, wipe up spills immediately. Hard water stains, soap film, and mildew can be removed with a commercial mildew-removal product.

ADHESIVE AND GROUT

Traditionally, tile was set in a thick bed of mortar. This may still be the best choice if you have a floor with a lot of bounce. In this case you may want a "floating floor" installed using layers of building paper and wire mesh and a thick mortar bed; this job is best left to the professional. In other cases, by far the simplest approach is to use an adhesive to bond the tile to the floor. After the tile is set, the joints between the tiles are filled with grout to keep out dirt and water and also to bond the tiles together. Always check with your tile dealer to make sure your adhesive and grout are compatible with the type of tile you're using.

ADHESIVE

The best adhesives for residential use are mastics or cement-base thin-sets. The characteristics of the two types of adhesive are discussed below. Epoxy adhesives are used in commercial applications where the tile will be subjected to very heavy traffic.

Mastics (organic adhesives): These adhesives are referred to as "organic" because they were originally a natural rubber product. Mastics come in a can in liquid form. Solvent-base mastics should be avoided because of the toxic and flammable fumes; they have largely been replaced by latex (water-base) products. Mastics are cheap and easy to apply; however, latex mastics are not very water-resistant and should only be used in rooms that are not going to be subject to a lot of moisture. Mastics are best used over plywood and are applied with a notched trowel.

Thin-sets (cement-base adhesives): Like the traditional mortar used for setting tile, thin-sets contain dry portland cement and sand. However, only a thin layer is needed, and, like mastics, thin-sets are applied with a notched trowel. Thin-sets are available with a latex or a polymer additive. The latex additive is in a liquid form, while the polymer additive is a powder mixed with the thin-set. Modified thin-sets, particularly those with a latex additive, offer the advantage of being more water-resistant than mastics, and being usable over a greater variety of backings. In addition to plywood, suitable backings for thin-sets are cement backerboard, concrete, and masonry, including old tile and brick.

GROUT

The two basic categories of grout are cement-base grout and epoxy grout. The type you should use will depend on the type of tile that you're installing, where it's located, the adhesive it's set in, and the width of the joints that you're leaving between the tiles. Ask your tile supplier to recommend the specific type of grout for your project. Silicone rubber caulk is used where a flexible joint is required.

Cement-base grouts: Used by both amateur and professional tile installers for finishing most ceramic tile floors, these grouts are made with a base of portland cement modified with a latex or polymer additive to provide characteristics such as water-resistance, hardness, flexibility, and uniformity. These grouts, particularly those with a latex additive are fairly resistant to water and stains, but are not completely water- or stain-proof.

Epoxy-base grouts: The main advantage of these grouts is that they are highly resistant to both water and chemicals. They are the best choice for a kitchen, workroom, or darkroom floor. However, they are also more expensive than cement-base grouts, and are more difficult to work with. Epoxy-base grouts must be mixed carefully at the site and are extremely difficult to clean up once they've begun to harden.

Silicone rubber caulk: Not strictly a grout, silicone rubber caulk stays permanently flexible, repels water, and resists mildew. It is often used to seal the seams around bathtubs and other bathroom fixtures. Although silicone rubber caulk is too soft for use between floor tiles, it should be used for filling the joints between a rigid tile floor and an adjoining floor of wood or resilient material, and for filling the joint around the perimeter of a tile floor. The flexibility of the caulk will allow the surfaces to move independently.

 PLAY IT SAFE

WORKING WITH EPOXY GROUT

Follow all safety precautions given on the package. Epoxy grout can be irritating to the skin, the fumes can irritate the nose and throat, and it can cause damage if it gets splashed in the eyes. Wear eye protection as well as rubber gloves and long sleeves to avoid contact with the skin. Work in a well-ventilated area, and if you find you are still sensitive to the fumes, try wearing a respirator (check with your safety equipment supplier for the right filter and cartridge). Alternatively, you can switch adhesives or hire a contractor to complete the job.

TOOLS

To lay out your project, you'll need the tools shown on page 30. You'll also need some basic carpentry tools and skills to prepare the backing for your tile; these skills are not covered in this book. In addition, you'll need some of the specialized tools shown below for marking, cutting, and installing ceramic tile. To cut very heavy ceramic tile, consider renting a wet saw from your tile dealer. A saber saw with a carbide-grit edge blade can also be used to cut ceramic tile.

Before you begin, make sure you have the necessary safety equipment:

• Eye protection: Wear when using any striking tool, and when working with a material that could splash, such as adhesive or grout.
• Work gloves: Wear when working with sharp or rough materials.
• Rubber gloves: Wear when using caustic products, such as epoxy grout.
• Dust mask: Wear when working with dusty materials, such as dry adhesive or grout.
• Respirator: Wear when using any products that give off toxic fumes, such as epoxy grout.

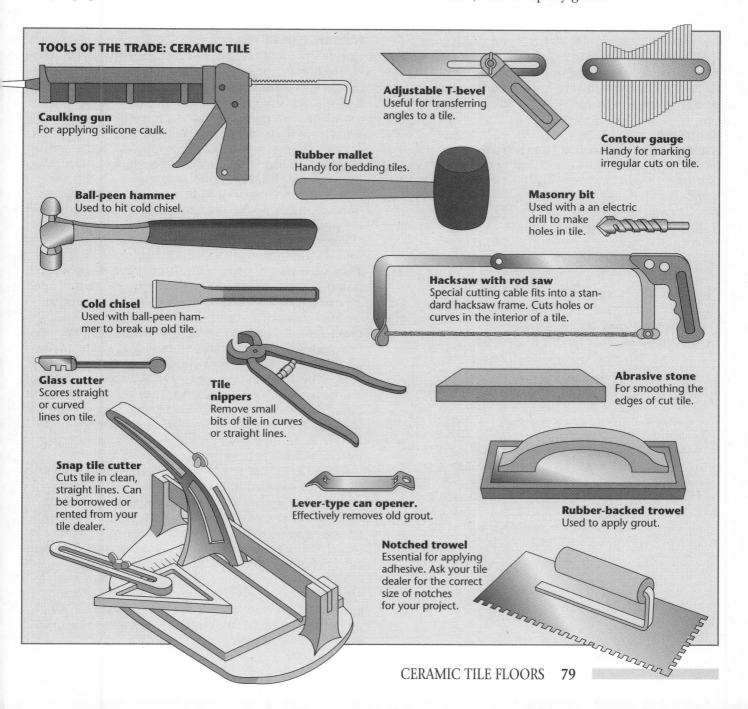

TOOLS OF THE TRADE: CERAMIC TILE

Caulking gun
For applying silicone caulk.

Adjustable T-bevel
Useful for transferring angles to a tile.

Contour gauge
Handy for marking irregular cuts on tile.

Rubber mallet
Handy for bedding tiles.

Masonry bit
Used with a an electric drill to make holes in tile.

Ball-peen hammer
Used to hit cold chisel.

Hacksaw with rod saw
Special cutting cable fits into a standard hacksaw frame. Cuts holes or curves in the interior of a tile.

Cold chisel
Used with ball-peen hammer to break up old tile.

Glass cutter
Scores straight or curved lines on tile.

Tile nippers
Remove small bits of tile in curves or straight lines.

Abrasive stone
For smoothing the edges of cut tile.

Snap tile cutter
Cuts tile in clean, straight lines. Can be borrowed or rented from your tile dealer.

Lever-type can opener.
Effectively removes old grout.

Rubber-backed trowel
Used to apply grout.

Notched trowel
Essential for applying adhesive. Ask your tile dealer for the correct size of notches for your project.

SUBFLOOR PREPARATION

Before you begin, inspect the subfloor from above and below for any problems that may require repairs, as described on page 21. Many existing floors can serve as a backing for ceramic tile, as can a concrete slab or a plywood subfloor. You'll find instructions below on how to prepare each type of surface.

CONCRETE SLAB

Remove grease and oil stains from the slab with a chemical garage floor cleaner available at most auto supply stores. Remove old paint or glossy sealers with a drum sander fitted with No. 4 or 5 open-cut sandpaper. For information on renting and operating drum sanders, refer to page 54. Vacuum or sweep the floor as you go.

Fill any holes, low areas, or cracks in the slab with a concrete patching compound or with thin-set adhesive. Finally, scour the slab with a stiff bristle or wire brush. Sweep or vacuum up all loose particles.

WOOD SUBFLOOR

Make sure that all panels or boards of an existing subfloor are securely fastened. A wood subfloor, either plywood or board, should be covered with a plywood underlayment. The combined thickness of subfloor and underlayment should be 1¼ inch, with the underlayment at least ⅜ inch. If the combination is not thick enough, cover it with a layer of ⅜-inch exterior-grade plywood.

PLAY IT SAFE

CAUTION: YOUR OLD RESILIENT FLOORING MAY CONTAIN ASBESTOS

If your resilient flooring was installed before 1986, it may contain asbestos. Asbestos poses a health hazard if the fibers are released into the air. The only way to find out for sure whether your floor contains asbestos is to send a sample to a private lab for testing. If you don't know whether your floor contains asbestos, DO NOT sand it. Generally, the simplest approach is to cover the old floor with plywood; if you want to remove the old flooring, check with your local Environmental Protection Agency office or state health department to find out whether this is allowed. If you do remove the flooring, safety precautions must be followed; these are available from the Resilient Floor Covering Institute in Rockville, Maryland.

When putting down plywood underlayment, make sure that the joints do not fall directly over those in the layer below. Use 4x8-foot exterior-grade panels and stagger the joints; leave a ⅛-inch gap between the panels. Use ring-shank nails twice as long as the new plywood is thick, and space the nails 6 inches apart along the edges and 8 inches apart in the field. Nail directly into the joists where possible.

You may want to consider using cement backerboard instead of plywood as an underlayment. Cement backerboard is moisture-resistant and will not curl. Follow the manufacturer's instructions for installation.

EXISTING FLOORS

Removing old flooring before installing new ceramic tile makes it easier to examine the subfloor and make any necessary repairs; it also means there's a better chance that the new floor will be level with the floors in adjacent rooms. However, you can lay the new tile directly over certain old floor surfaces.

You'll have to take up the shoe molding before putting in the new floor. Gently pry the molding away from the wall, using a thin-blade prybar or a butt chisel. Baseboards do not need to be removed unless there is no shoe molding. As you remove the molding, number the pieces with a pencil so you can easily replace them.

Wood floors: Wood strip and plank flooring are not smooth enough to serve as backings for ceramic tile. They should be covered with ¼-inch exterior plywood, fastened as described above for plywood underlayment.

Resilient floors: If it's level and in good repair, well-bonded resilient flooring can be covered with tile. To make minor repairs to resilient flooring, see page 72. Cushioned resilient flooring is too springy to be used as a base for ceramic tile and must be removed or covered over. Other resilient flooring that is badly damaged should also be removed or covered over. To cover an old resilient floor, use ¼-inch exterior-grade plywood panels. Fasten them as described above for plywood underlayment. NOTE: Your old resilient floor may contain asbestos; refer to the safety information at left.

Old ceramic tile: Tile can be applied over old ceramic tile if the old tile is in good condition, clean, and well-bonded. Pry up any loose tiles and check the backing. If there is evidence of water damage, the tiles and backing may have to be removed and the moisture problem corrected; consult a professional. If there is no problem with the backing, replace the loose tiles *(page 87)*.

To prepare the old tile, clean it with a commercial degreasing agent. Apply the adhesive to a small area and examine it the next day. If it hasn't bonded well, roughen the surface of the tile with an abrasive disc mounted on an electric drill. Wear a dust mask for the job.

LAYING CERAMIC TILE

Before you begin, make sure you have everything you'll need for the job. Open and inspect the cartons of tile to make sure you were given the right type and color, and that you have enough to cover the floor. Read the directions on the adhesive container to be sure it is the proper adhesive in the quantity you'll need. If the tile is dusty, you should wash and dry it before commencing the installation. Even a little dust can prevent adhesives from forming a strong bond.

The key to laying a floor with straight rows of tile is to lay out proper working lines independent of the walls. You can begin either at the center of the room or at one wall. If you're tiling from the center, begin by laying out working lines that intersect at a right angle at the approximate center of the room. This is the best method to use if the room is badly out of square or if you've chosen tile with a definite pattern or design. This method will likely mean cutting tiles on all four walls, but it will allow you to have balanced cut tiles on opposite walls. For instructions on laying down working lines in the center of a room, see page 43.

Start at the wall *(page 82)* only if two adjoining straight walls meet at an exact 90° angle. This method will mean cutting tiles on only two adjacent walls.

MAKING CLEAN CUTS

Mark 90° cuts on tile using a combination square. Angles can be transferred from a wall to a tile with an adjustable T-bevel, and irregular shapes can be transferred with a contour gauge.

Always wear eye protection when cutting tile. To make straight cuts near the middle of the tile, you can use a snap tile cutter or a glass cutter. First, mark the cut with a pencil. If you're using a snap tile cutter, score the finished surface along the pencil line with the cutter *(right)*; then press down on the handle to break the tile. Or, you can score the face with a glass cutter and straightedge *(below, left)*. To break the tile, place the scored line over a dowel or nail and press down evenly on both edges *(below, right)*. To make irregular cuts, or to remove a small amount of tile near the edge, score the tile with a glass cutter, and then use tile nippers to nibble up to the scored line. For interior cuts, first drill a hole with an electric drill and masonry bit; then make the cut with a rod saw. Smooth cut edges with a special abrasive stone, or with a whetstone.

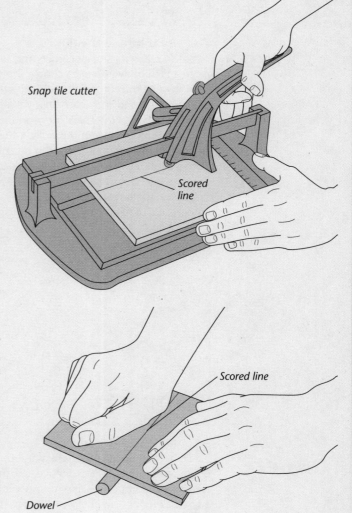

Snap tile cutter

Scored line

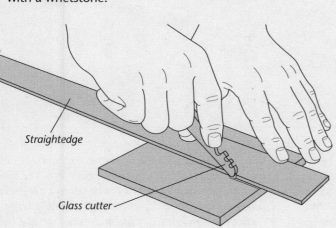

Straightedge

Glass cutter

Scored line

Dowel

TOOLKIT
• Chalk line
• Tape measure
• Carpenter's square
• Claw hammer
• Tile stick (optional)

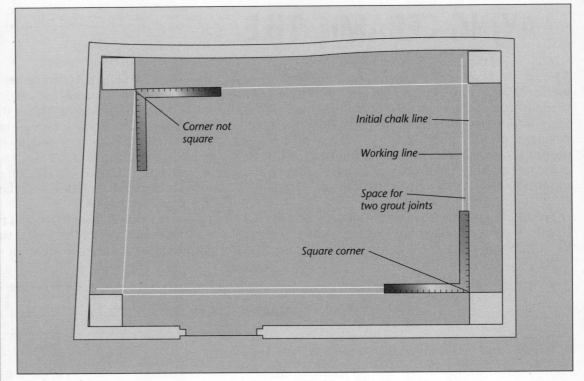

Corner not square

Initial chalk line

Working line

Space for two grout joints

Square corner

1 Establishing the lines

To check for square corners and straight walls, place a tile tightly into each corner. Stretch a chalk line between the corners of each pair of tiles, pull the line tight, and snap each line. Variations in the distance between the chalk lines and the walls will show any irregularities in the walls. You can ignore small variations—about the width of a grout joint. With a carpenter's square, check for square where the lines intersect in each corner of the room.

You can start at any straight wall adjoining a square corner. Snap a new chalk line parallel to the first line

and approximately two grout joint widths away from it toward the center of the room, as shown above. Lay a similar line, at a right angle to the first, along an adjoining wall.

Next, nail a wooden batten (a 1x2 or 1x3) along each of the new working lines. The two battens should form a right angle; if they don't, check your measurements and adjust the working lines accordingly. If you're working over a concrete slab, it will be difficult to fasten the battens; you may want to rely on chalk lines without battens.

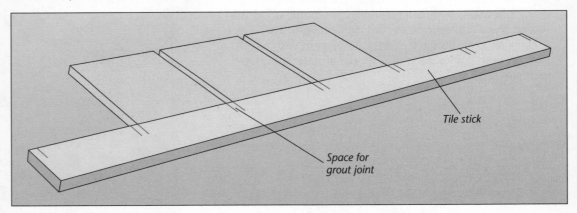

Tile stick

Space for grout joint

2 Making a dry run

It's a good idea to make a dry run before you begin setting the tiles in adhesive. Lay the tiles out, allowing the proper spacing for grout joints. Using a tile stick *(above)* will help you achieve uniform spacing—mark a board for the width of the tile you're using, and for the width of the grout joints.

You can also use plastic or wooden spacers. The dry run will help you determine the best layout of tiles and keep the number of tiles to be cut to a minimum; you may find that a slight reduction or enlargement of grout joint widths will make it possible to set a row of tiles without any cutting.

Laying ceramic tile

TOOLKIT
- Notched trowel
- Carpenter's square
- Hammer or rubber mallet
- Prybar

1 Spreading adhesive

Start spreading a strip of adhesive on the floor along one of the battens, using the smooth edge of the notched trowel; then go back over the area with the notched edge. Work back and forth in a crosshatch pattern. To double-check that you have the right trowel, spread some adhesive and bed a tile. Then, pull up the tile—if less than 95% of the back of the tile is covered with adhesive, the notches are too small; if adhesive squeezes up around the edges of the tile, the notches are too big. Don't cover more area than you can comfortably tile within the open time of the adhesive. (Open time is the amount of time you have before the adhesive begins to set; it varies with the adhesive.) As you become more proficient, you may be able to spread the adhesive over a larger area.

If you're working from the center of the room, begin spreading adhesive on one side of a working line as shown on page 70.

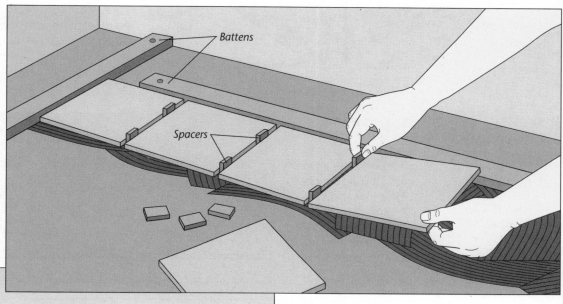

Battens

Spacers

Second row

2 Setting the tiles

Using a gentle twisting motion, place the first tile in the corner formed by the two battens. (For large tiles, such as 10"x10", spread some adhesive on the back of the tile before setting it.) With the same motion, place a second tile alongside the first. To establish the proper width for the grout joint, use uniform spacers cut from wood, or molded plastic spacers (available from your tile dealer); set plastic spacers on end along the edge of each tile, and not in the corners, so they'll be easy to remove; take them out when the adhesive begins to set.

If you're working from the center of the room, lay tiles in the sequence shown on page 44. If you've started from a wall, set the tiles in continuous rows, beginning each row at the same end of the room, as shown at left.

As you work, clean off any adhesive from the surface of the tiles. From time to time, use a carpenter's square to check your work. If any tiles are out of line, wiggle them back into position. If you need to work over the newly set tiles, use a sheet of plywood to distribute your weight evenly.

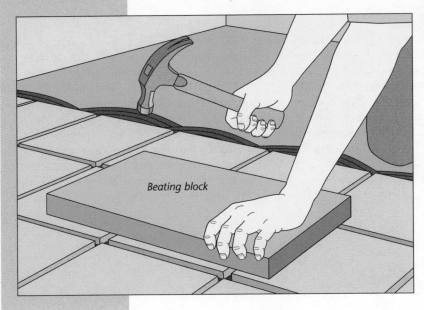

Beating block

3 Bedding the tiles and finishing the job

When you finish a section of the floor, bed the tiles by tapping each one with a rubber mallet, or by tapping a beating block—a piece of wood that spans more than one tile—with a hammer (left).

When you're ready to install the border tiles, remove the battens. Then mark the border tiles as shown on page 46. Cut the tile as explained on page 81.

When all the tile has been laid, make sure you've removed all the spacers; then clean the tile surface so it's completely free of adhesive. Use a damp cloth to remove any adhesive that has pushed up between the joints.

Allow the adhesive to set properly before applying grout; check the manufacturer's directions for setting time. Keep all unnecessary traffic off the floor until the tiles have been grouted.

Grouting and finishing

TOOLKIT
- Rubber-backed trowel
- Claw hammer
- Tool for applying sealer (optional)

1 Applying the grout

Pour about 1 cup of grout onto the floor at a time. Spread it using a rubber-backed trowel held at a 30° angle (right). Force grout into the joints so they're completely filled; check that there are no air pockets.

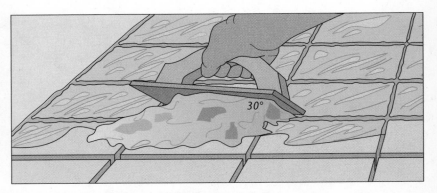

30°

2 Removing the excess

Scrape off the excess grout with a rubber-backed trowel held at 45°. Work diagonally to avoid disturbing the grout joints. Clean off your trowel frequently in a bucket of water.

45°

3 Cleaning and sealing

Wipe the tiles with a damp sponge, using a circular motion, to remove any remaining grout until the joints are smooth and level with the tiles. Rinse and wring out the sponge frequently. Let the grout dry until a haze appears on the tiles. Then, polish the tiles with a soft cloth.

Let the grout cure for two weeks. Then, if you've used a cement-base grout and the floor is likely to be subjected to staining or to standing water, you should seal the grout. Except for porcelain tile, unglazed tile should also be sealed. Follow the manufacturer's instructions for applying grout or tile sealer. If you're sealing the grout only, make sure that you wipe any sealer off the tiles before it hardens.

If you removed any shoe molding, replace it by driving nails through the molding directly into the baseboards.

REPLACING A CERAMIC TILE

The qualities that make ceramic tile a desirable flooring material—its hard surface and rigidity—also make it susceptible to cracking. A heavy object dropped on tile can easily crack or chip it. Structural problems, such as cracks in a concrete slab or the settling that takes place in a frame house, can lead to a series of cracks in a tile floor. Structural shifting may leave individual tiles intact while opening gaps between tiles or causing grout to deteriorate. Sometimes the grout around perfectly good tiles may need to be replaced when it becomes stained or cracked.

Although replacing a single ceramic tile or a few tiles isn't difficult, finding a match between replacement tiles and an old floor can be a problem. Colors may vary from one firing to another, even when the manufacturer uses identical glazes. If you don't have matching tiles left over from the original installation, you may be able to find an acceptable substitute by checking with several flooring or specialty tile stores. Should you find nothing—a likelihood particularly if you're replacing a pattern that's been discontinued—consider using complementary colors as replacements.

TOOLS AND SUPPLIES

To remove damaged ceramic tiles, you'll need the following tools: a glass cutter, a common lever-type can opener, a hammer, a nailset or center punch, a putty knife, a hand-drilling hammer or ball-peen hammer, and a small cold chisel. You may want a rubber-backed trowel for applying new grout.

To replace tiles of odd shapes or around bathroom fixtures or similar obstructions, you'll need a pair of tile nippers, as well as an abrasive stone. These and other specialty tools needed for working with tile are illustrated on page 79.

Finally, you'll need adhesive and grout to cement the new tile in place. Your flooring materials supplier will be able to help you select the correct adhesive for the job and provide a compatible grout.

CAUTION: Always wear eye protection when you're chipping out old ceramic tile.

Replacing grout

Removing the old grout and regrouting
Remove the old grout with a lever-type can opener, as shown below. Then, scrub the joint surfaces with scouring powder or tile cleaner; rinse well. Then apply the new grout (*page 87*).

Replacing a ceramic tile

TOOLKIT
- Lever-type can opener
- Claw hammer
- Nailset or center punch
- Glass cutter
- Straightedge
- Cold chisel
- Ball-peen or hand-drilling hammer
- Butt chisel (optional)
- Putty knife
- Rubber-backed trowel or rubber squeegee (optional)

1 ▷ Removing the grout
Using a lever-type can opener, remove the old grout from the joints around the damaged tile.

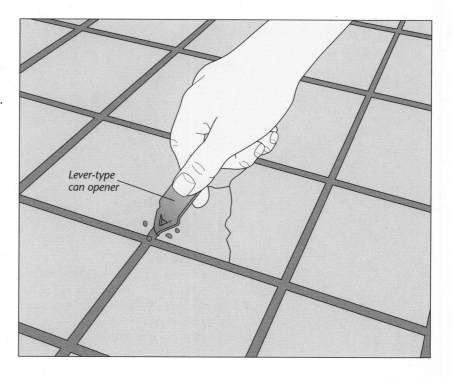

Lever-type can opener

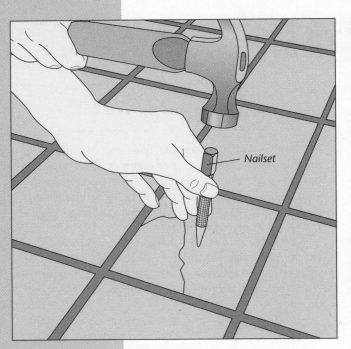

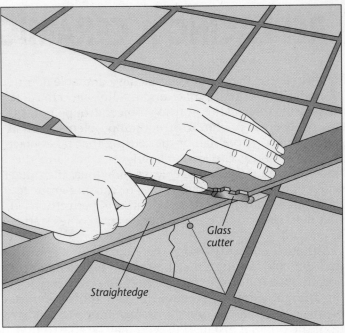

Nailset

Glass cutter

Straightedge

2 **Scoring the tile**
Punch a hole through the center of the damaged tile using a hammer and nailset *(above, left)* or a center punch. Pound gently so as not to damage the subfloor. Then, use a glass cutter and straightedge to score a deep X across the face of the damaged tile through the center of the hole *(above, right)*.

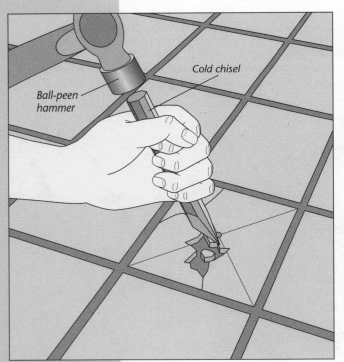

Ball-peen hammer

Cold chisel

Putty knife

3 **Removing the tile**
Starting at the center, chip out the old tile (and any remaining grout) using a cold chisel and a ball-peen hammer *(above, left)* or hand-drilling hammer; use light, rapid blows. Remove any remaining adhesive or grout from the surface using a putty knife *(above, right)*, cold chisel, or butt chisel. Use sandpaper to smooth rough spots and the edges or surrounding tiles. Build up the surface, if necessary, with a cement-base patching compound.

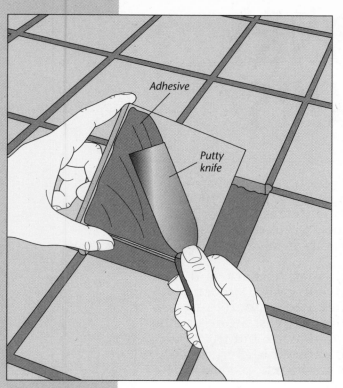

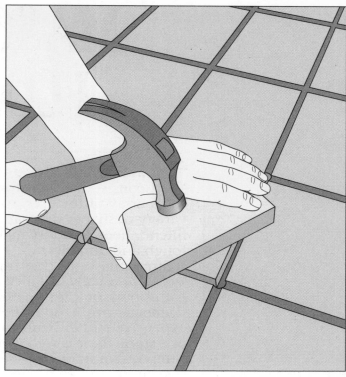

4 Installing a replacement tile

Apply adhesive to the back of the replacement tile or to the subfloor, using a putty knife *(above, left)*. Then, using a hammer and wood block, tap the tile gently into place *(above, right)*. Allow the adhesive to set for at least 24 hours before grouting. Keep traffic off the new tile.

5 ▶ Applying grout

Apply the new grout using a rubber-backed trowel, a rubber squeegee or a putty knife. Use your finger to smooth the grout joint. Clean off any excess with a damp sponge, and once the grout has dried, polish the tile with a soft cloth.

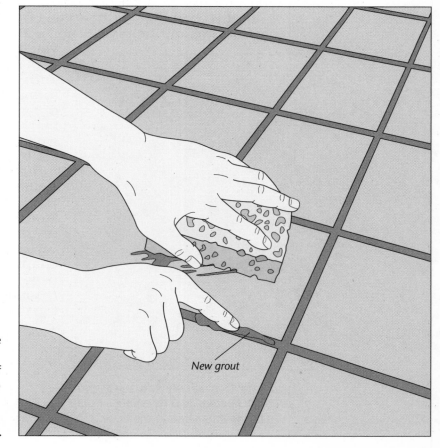

MASONRY FLOORS

Masonry materials are products of the earth and consist of either stone or clay. They come in an array of textures, subtle shades of color, and a variety of shapes that can be used to create a striking interior floor. Masonry materials are also exceptionally tough and durable.

Stone masonry flooring—marble, slate, and flagstone—is usually installed in entryways, kitchens, and bathrooms, or it is sometimes used as flooring around fireplaces and wood-burning stoves. Stone can be expensive and difficult to work with, but the beauty of a stone floor is well worth the effort. Terrazzo tile, composed of stone chips set in concrete, is a less expensive alternative to natural stone.

Brick, a relatively inexpensive type of masonry, is becoming an increasingly popular choice for interior floors. It can be laid in a variety of attractive patterns. The mass and heat-retaining property of brick make it ideal for passive solar home designs.

In this chapter we'll show you how to select masonry flooring and help you choose the specialty tools you'll need for working with masonry *(page 90)*. Turn to page 91 for information on preparing a base for masonry. Starting on page 92, we'll show you how to set brick with sanded joints and brick or rough-hewn stone in a thick base of mortar. Stone tiles are installed in the same way as ceramic tiles; turn to the chapter that starts on page 76. For tips on repairing masonry floors, turn to page 96.

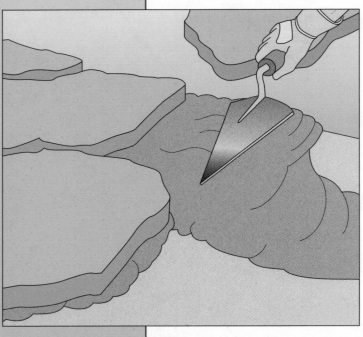

Mortar for rough-hewn stone is spread stone by stone to compensate for the differing thicknesses of the stones.

SELECTING MASONRY MATERIALS

Because of their weight, masonry materials are difficult to transport and their availability will vary with the locality. Unless you're willing to go to great expense, you may be limited in your selection. Masonry materials include clay bricks, manufactured terrazzo tiles, and natural stone.

Brick: Bricks intended for walls often have large holes in them and are not suitable for floors. Bricks intended specifically for floors, whether inside or outside, are called pavers. These are available in regular or slightly less-than-regular thicknesses, or as splits—half as thick as regular bricks. Since splits are much lighter than full bricks, the floor will require less reinforcement to support them.

Stone: For interior flooring, choose stone that is dense, with a hard surface. Stone is available either in rough-hewn pieces, or in uniform tiles.

Rough-hewn stone means pieces of irregular shape and thickness. Flagstone is made of any type of stone that can be easily cut or split into thin slabs, such as sandstone. Other types of rough-hewn stone include granite and slate. Rough-hewn stone is always set in a thick mortar base. The mortar compensates for variations in thickness and provides a solid base for the irregular undersides of the stones. Keep in mind that the thickness of these stones (often around 2 inches), in addition to the mortar bed, will raise the height of your floor considerably. Also, the weight of rough-hewn stone will require a very strong subfloor.

Marble or slate can be ordered in thin tiles of uniform thickness. They usually come in the form of square tiles, but are also available in irregular shapes. Marble or slate tiles can be laid in thin-set adhesive. Follow the directions given in the chapter on ceramic tile *(page 76)*, making sure you have selected an adhesive that is appropriate for stone. Stone tiles set in adhesive are a good choice over a wood floor, as they make for a much lighter floor than flagstone flooring set in a full bed of mortar. The subfloor requirements are the same as those for ceramic tile.

Terrazzo tile: These tiles are manufactured by setting chips of marble or onyx in concrete and then polishing the surface. Although commonly made in large slabs, terrazzo is also available in 12-inch-square tiles. These can be installed with thin-set adhesives in the same way as ceramic tile.

MIXING MORTAR

Mortar typically contains portland cement, masonry sand, hydrated lime, and water. Portland cement is sold dry in large bags. Masonry sand can be purchased—never use beach sand. Hydrated lime is available dry in sacks; it is not the same as garden lime. Always use drinkable water for mixing mortar.

The best mortar for laying brick floors is Type S, sold premixed in bags. To mix it yourself, use one part portland cement, $1/2$ part lime, and $4^{1}/_{2}$ parts sand. Mortar for stone should not contain lime, since it may stain the stone. Either buy premixed mortar without lime or mix the mortar yourself. The best proportions are one part portland cement to three parts sand.

To mix mortar, you can use a wheelbarrow or mortar box and a mortar hoe. Mix the dry ingredients first and then add the liquid gradually until the correct consistency is achieved. The mortar should be smooth and uniform. It should spread easily, like soft butter, without slumping or losing its shape. Another option is a "walk-along" mixer. Add half the water and sand and walk the mixer along to blend them. Then add the rest of the ingredients and continue to blend.

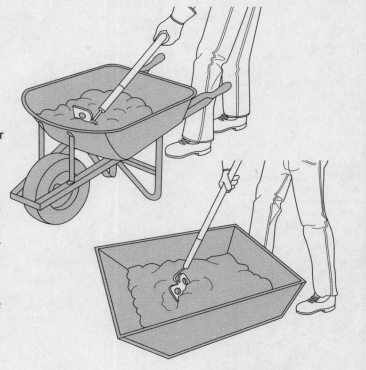

TOOLS

The specific type of masonry floor you're installing determines the tools and supplies you'll need for the job. The specialty masonry tools you'll need are shown below, as well as the tools you'll need to lay out your project. You'll also need a square shovel to measure mortar ingredients, and you can use a wheelbarrow instead of a mortar box for mixing mortar. A soft bench brush is best for brushing sand into the joints of a brick floor. For repairs, you'll need a cold chisel.

Before you begin, make sure you have basic safety equipment on hand. Mortar, dry or wet, is caustic—wear long sleeves and work gloves. Wear safety goggles or safety glasses when cutting brick or stone, and a dust mask when working with dry mortar ingredients.

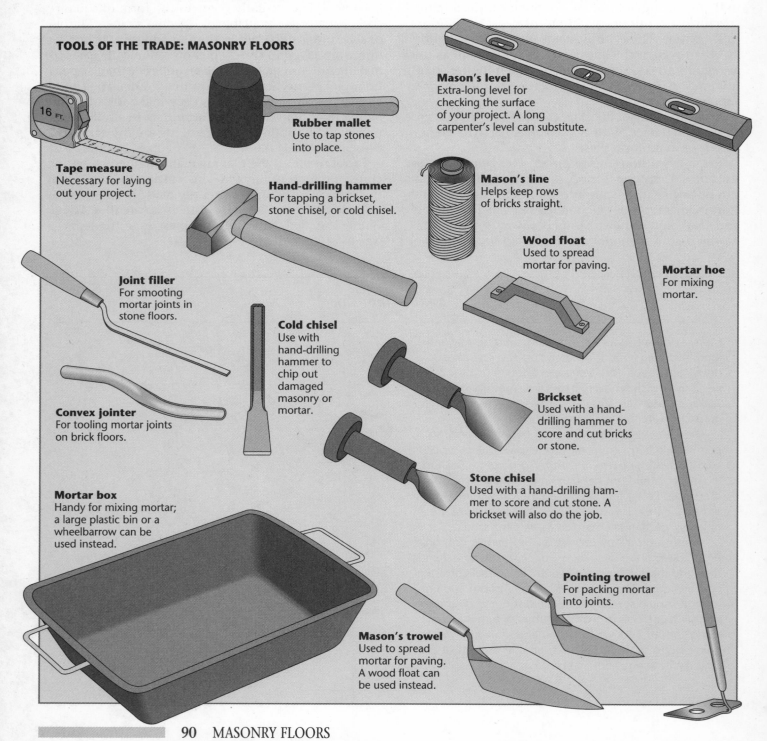

TOOLS OF THE TRADE: MASONRY FLOORS

Mason's level
Extra-long level for checking the surface of your project. A long carpenter's level can substitute.

Rubber mallet
Use to tap stones into place.

Tape measure
Necessary for laying out your project.

16 FT.

Hand-drilling hammer
For tapping a brickset, stone chisel, or cold chisel.

Mason's line
Helps keep rows of bricks straight.

Wood float
Used to spread mortar for paving.

Mortar hoe
For mixing mortar.

Joint filler
For smooting mortar joints in stone floors.

Cold chisel
Use with hand-drilling hammer to chip out damaged masonry or mortar.

Convex jointer
For tooling mortar joints on brick floors.

Brickset
Used with a hand-drilling hammer to score and cut bricks or stone.

Stone chisel
Used with a hand-drilling hammer to score and cut stone. A brickset will also do the job.

Mortar box
Handy for mixing mortar; a large plastic bin or a wheelbarrow can be used instead.

Pointing trowel
For packing mortar into joints.

Mason's trowel
Used to spread mortar for paving. A wood float can be used instead.

SURFACE PREPARATION

asonry flooring demands a firm, unyielding base. It can be laid over either a ground-level concrete slab or a suspended wood floor. However, putting masonry over a suspended floor in a frame home can be complicated because of the weight of masonry. If masonry flooring was planned as part of the original building, allowances will have been made for the extra weight and the support system strengthened accordingly. For example, to support a brick floor, the joists should be spaced about 25% closer together to handle the extra weight (about 15 pounds per square inch); joists should therefore have 12-inch instead of 16-inch centers. Split bricks require less support. Laying rough-hewn stone over a wood floor can be particularly problematic because of the great weight and inflexibility of the floor. If you're installing stone tiles, the requirements are the same as for ceramic tile.

If brick or stone flooring is being laid in an older home as part of a remodeling project, the old floor will likely need reinforcing. Requirements will vary according to the material you're installing and local building codes. Always consult a building design professional before putting masonry flooring over any subfloor other than a ground-level concrete slab. If you're setting masonry over a wood floor, the floor will have to be protected by a moisture barrier of asphalt felt.

If you're planning to install masonry flooring in a mortar bed over a concrete slab, make sure the slab is free from dust or grease that could affect its bonding with the mortar. Grease and oil can be removed with a floor cleaner available at auto supply stores. Other dirt can be removed using a stiff bristle brush, and, if necessary, detergent. Dampness should present no problem, as long as the slab has cured fully. Allow 30 to 60 days for a new slab to cure.

If you're setting bricks with sanded joints, you should fill holes and dips in a concrete slab with a concrete patching compound. If you're spreading a mortar bed, the bed will take care of most minor irregularities.

If there is already an old floor in place, the best approach, where possible, is to expose the subfloor. Leaving the old floor there will increase the weight of the system and take away from the headroom. If you want to leave an old wood floor in place, cover it with asphalt felt as you would a wood subfloor. A mortar bed can be placed over old ceramic tile if the tile is first covered with a bond coat of thin-set adhesive (check with a tile supplier). NOTE: Removing old resilient flooring can be dangerous because it may contain asbestos; see the safety information on page 72. The simplest approach may be to cover the old flooring with 1/4-inch exterior-grade plywood.

QUICK FIX

BRIDGING CRACKS IN CONCRETE

If you're laying masonry in a mortar bed over a concrete slab, bridge hairline cracks to prevent the masonry from cracking. Commercial crack-suppression membranes come in kits with a fabric tape and liquid rubber; the membrane should be wider than the units being installed. Large cracks in concrete may indicate a structural problem, particularly if there is to be movement—if one side of the crack is higher than another. In such cases, consult a professional.

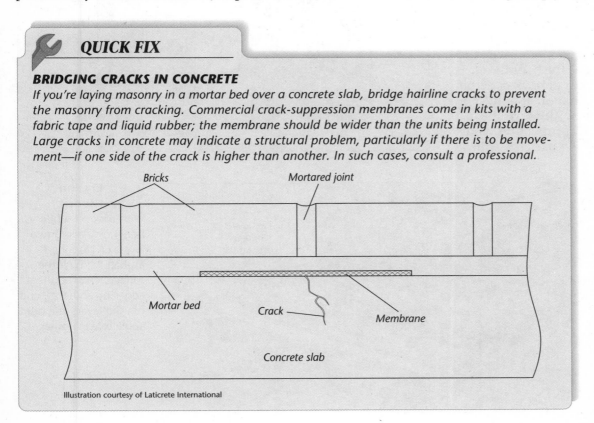

Illustration courtesy of Laticrete International

INSTALLING A BRICK FLOOR

Bricks can be laid in a number of ways. The simplest is to set them directly on the subfloor with butt joints; sand is then brushed into the tiny gaps between the bricks. This is the most common method used for interior floors, and by far the easiest for the amateur; it produces a beautiful brick floor with minimal fuss. The other method described in this section is to set the brick in a bed of mortar with mortared joints. Although this method calls for more skill, the result is strong and handsome, and often worth the effort. A third alternative is to set the bricks in a mortar bed, but with butted joints; sand is then forced in between the bricks. This offers the solidity of a mortar bed without the complication of mortar joints. Finally, if you're working with very thin bricks—$1/2$-inch, for example—you may want to consider laying a brick floor in thin-set adhesive; simply follow the instructions for ceramic tile *(page 76)*.

Bricks laid over a wood subfloor should not exceed $1^5/8$ inches in thickness; over concrete slabs, any thickness will do. If you're laying bricks over a wood subfloor, you should protect the subfloor with asphalt felt. If you're using a mortar bed, you may want to put down a layer of wire mesh before laying the mortar to keep the mortar from cracking as the floor flexes.

Over the years, dozens of brick patterns have evolved; the illustration at right shows the most common ones. Although the most common pattern is running bond, you could choose any of the patterns shown.

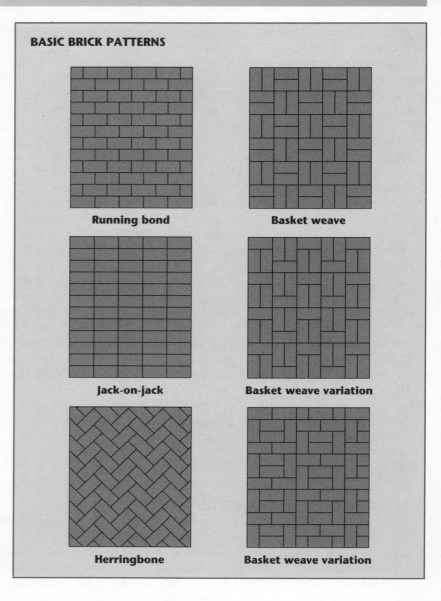

BASIC BRICK PATTERNS

Running bond

Basket weave

Jack-on-jack

Basket weave variation

Herringbone

Basket weave variation

Laying a mortarless brick floor

TOOLKIT
• Tape measure
• Mason's line
• Mason's level
• Hand-drilling hammer
• Brickset
• Bench brush

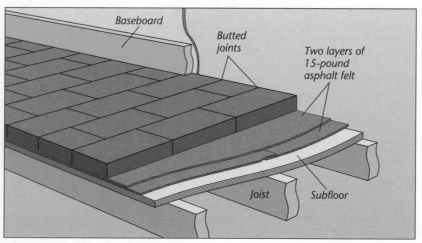

Baseboard

Butted joints

Two layers of 15-pound asphalt felt

Joist

Subfloor

1 Placing the felt
With a wood subfloor, begin by placing two layers of 15-pound asphalt felt over the surface. Overlap the edges by about 2", and run the layers at a right angle to each other.

2 Laying the bricks and spreading the sand

If you have a truly straight wall, you can use it as a guide. Otherwise, stretch a mason's line a little less than one brick length or width away from the starting wall, depending on the pattern you're using; wrap the ends of the line around loose bricks. Lay the first course of bricks up to this line. After the rest of the bricks are in place, you'll go back and cut bricks to fit between the first course and the wall. Reset the line for each course or two. You can adjust joints between bricks slightly to keep the courses straight; just remember that a tightly butted joint is ideal.

Check the floor frequently for level. Because you can walk on the floor as it's laid, the starting and ending points are not important. Cut bricks to fit at the ends of each row as you go along.

When you've laid all the bricks, spread fine sand over the surface and sweep it diagonally into the joints with a soft brush. Repeat this sanding once or twice at intervals of a few days. Then, apply several coats of a masonry sealer, following the manufacturer's instructions. This will lock in the sand and also give a slight gloss to the floor.

ASK A PRO

HOW DO I CUT A BRICK?

Most bricks can be cut using a brickset. Score the brick on all sides by tapping the brickset with a hand-drilling hammer. Then make the cut with a sharp blow. Always wear eye protection. If you want very precise cuts, or if you're dealing with extremely hard bricks called "clinkers," you can rent a masonry saw or buy a special blade for your circular saw.

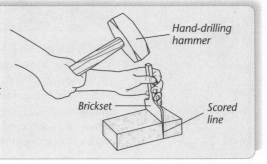

Hand-drilling hammer

Brickset

Scored line

Laying a mortared brick floor

TOOLKIT
- Tape measure
- Mason's line
- Mason's trowel or wood float
- Mason's level
- Hand-drilling hammer
- Brickset
- Pointing trowel
- Convex jointer, dowel, or other convex object

1 ▶ Setting the bricks

For a wood subfloor, lay asphalt felt as described opposite. Work toward an exit and situate your supply of bricks and mortar to avoid walking on the newly laid floor. Stretch a mason's line a little more than a brick length or width from the wall, depending on your pattern. Lay the first course of bricks dry, marking their positions on the floor; joints can be adjusted slightly to avoid cutting bricks.

If the bricks appear porous, soak them for a few minutes. Then, use a mason's trowel or wood float to spread an even layer of mortar over a small section of the floor. Try not to obscure your pencil marks. Set the first brick near the center of the line and work out in each direction. Use wood spacers: If the bricks are $3^1/2$" by $7^1/2$", the mortar joints should be $^1/2$"; bricks $3^5/8$" by $7^5/8$" should have $^3/8$" joints.

Check frequently for level. At walls or other obstructions, cut bricks to fit. Clean up mortar spills with wet burlap.

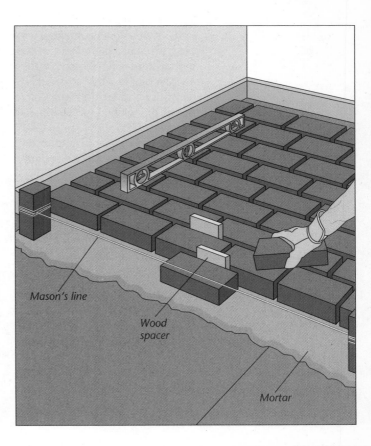

Mason's line

Wood spacer

Mortar

Pointing trowel

Convex jointer

2 Filling the joints

When the mortar has set for 24 hours, cut and set the row against the starting wall. Then you can begin filling the joints. Mix a small batch of mortar and, working carefully to minimize spills, use a small trowel to pack mortar into the joints. Alternatively, you can mix grout, which has the same ingredients as mortar but a soupier consistency. Using a can bent to form a spout, pour the grout into the joints. When the mortar or grout is just hard enough to keep a thumbprint, tool each joint with a convex jointer, a dowel, or other convex object; drag the tool along the joint to compact the mortar or grout.

3 Finishing the job

Wait several hours after the joints have been filled; then use a piece of damp burlap to remove mortar smears. If necessary, scrub the area with a solution of muriatic acid and water—1:9 for dark brick and 1:14 for light-colored brick; do not let the solution run over the edges of the bricks into the mortar joints. CAUTION: Always pour acid slowly into water, and never the reverse. Wear eye protection and rubber gloves.

Next, cover the new floor with a layer of plastic film (4-mil weight polyethylene is adequate) to keep the mortar damp while it cures. Leave the plastic film in place for about four days. After this, allow the floor to dry completely—this may take several weeks—before sealing the bricks. Any moisture trapped in the bricks when they're sealed can discolor the surface and spoil the seal.

CARING FOR YOUR MASONRY FLOOR

The porous surface of most masonry flooring must be sealed after the flooring has been installed. Special sealers are available for specific types of masonry. Most of these products are penetrating sealers that soak into the porous surface; depending on the flooring material, you may need to apply several coats to produce a hard surface. Your masonry supplier can recommend the appropriate product for the type of masonry flooring that you've installed. Follow the manufacturer's instructions to apply the sealer.

Once the surface has been sealed, masonry floors are very easy to maintain. The materials are rugged and durable and need only an occasional sweeping and washing with a damp sponge mop and mild detergent.

You may want to apply a light coat of wax now and then to bring out the character and texture of the flooring material. Check the label of the wax container to make sure the wax you've selected is compatible with the masonry material. Generally, you'll be looking for a water-base emulsion wax.

INSTALLING A STONE FLOOR

Rough-hewn natural stone is laid in a thick mortar bed for very practical reasons. The irregularity of the material makes it necessary to provide a cushion of mortar that will compensate for differences in thickness from piece to piece. Because rough-hewn masonry must be set with open joints, a sand base that might be practical for an outside patio is not workable as a base for an interior floor.

If you're working over a wood subfloor, you'll need to protect the floor with asphalt felt *(page 93)*. You may also want to put a layer of wire mesh over the felt to prevent the mortar from cracking as the floor moves.

Installing rough-hewn masonry units requires a bit of inventiveness, since each piece is a different shape and

thickness. Spread mortar as you go, varying the thickness of the bed to keep the surface level. Units must be trimmed to produce a satisfactory pattern with relatively uniform joints.

It's impossible to lay out a full floor of rough-hewn masonry in advance, but you can set out large sections dry before committing yourself to a mortar base. Distribute the larger pieces in a balanced pattern, and use smaller units or trimmed pieces to fill in the gaps. Once a section is bedded in mortar, move on to a new section, trying a dry fit as you did the first.

With rough-hewn masonry, it's essential to have a helper who can keep you supplied with mortar and help you move any large stones.

Setting stone

TOOLKIT
- Mason's trowel or wood float
- Rubber mallet (optional)
- Mason's level
- Stone chisel or brickset
- Hand-drilling hammer
- Pointing trowel
- Joint filler

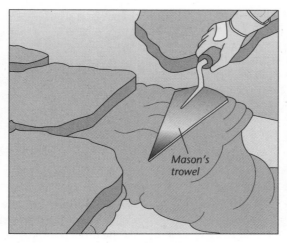

Mason's trowel

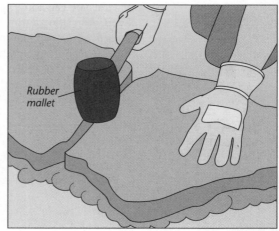

Rubber mallet

1 Placing and bedding the stones
Pick a corner of the room opposite your source of stone and mortar; you'll work toward your supply and, of course, an exit. If you're working over concrete, your slab should be slightly damp when you begin; sprinkle it with water, if necessary.

Use a trowel or wood float to spread mortar evenly over a small section of floor *(above, left)*. Don't

cover more area than is required to set two pieces of stone. Tap each stone level with a rubber mallet *(above, right)* or the handle of your trowel. Check frequently with a level to make sure each stone is level with its neighbors. Leave the joints between stones unfilled; you'll grout them later. Using a damp sponge, clean up mortar spills right away.

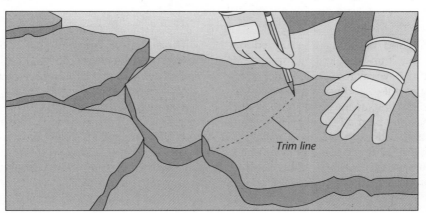

Trim line

2 Marking stones for cuts
As you go, trim the stone to fit along the walls and other obstacles. Lap the stone over its neighbor and mark the trim line *(left)*.

HOW DO I TRIM A STONE?

To cut a stone, first score it along a marked line, top and bottom, with a brickset or stone chisel and a hand-drilling hammer (right). Then, prop the edge to be cut on a piece of wood, and strike the scored line (far right).

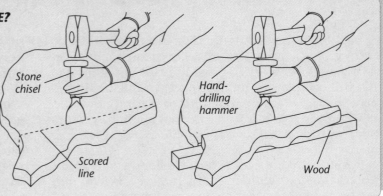

Stone chisel

Hand-drilling hammer

Scored line

Wood

3 ▶ **Filling the joints**

When the stone is laid, allow the mortar to set for 24 hours. Then, prepare mortar of the same proportions—3 parts sand and 1 part cement with no lime; you can add ½ part fireclay for workability, if you wish. Using a trowel, fill in the joints between units (right). Or, you can mix grout instead—the same ingredients but soupier than mortar; the consistency should be similar to a milk shake. Pour the grout into the joints from a bent can. Wipe up excess grout or mortar with a wet sponge.

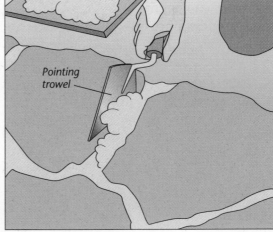

Pointing trowel

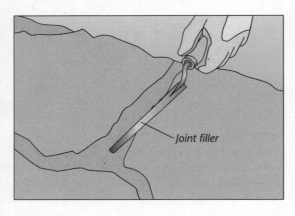

Joint filler

4 ◀ **Smoothing the joints and finishing the job**

As the grout or mortar sets, smooth the joints with a joint filler or trowel, leaving the grout or mortar flush with the surface of the stones.

When the floor has dried thoroughly, apply a sealer. Your building materials supplier can recommend one for the type of stone you've installed.

REPAIRING MASONRY FLOORS

Fortunately, masonry flooring materials are so rugged that they rarely need replacing. If units set in a mortar bed do need to be replaced, the only way to dislodge them is to put on a pair of safety goggles, and use a hand-drilling hammer and cold chisel to chip away at the stone or brick. A power drill with a masonry bit may come in handy, but in any case the job requires patience and a good deal of hard work. Bricks laid with sand joints can sometimes be pried out using a putty knife or old screwdriver. If this doesn't work, break up the damaged brick with a cold chisel.

The area from which the old flooring has been removed must be cleaned out and enough old mortar cleared away to make room for a new bed of mortar. A bonding agent can be applied to the surface before the mortar to improve the bond (ask your flooring materials dealer for a recommendation). Buy or prepare mortar as described on page 89. Apply the new mortar and install the new unit as described in this chapter. After the repair, the area should be resealed; consult your dealer for an appropriate product.

CARPETING

Carpeting gives a warm, welcoming look to your room, and provides comfort underfoot. In addition, it insulates and reduces noise. Carpeting is available in a range of colors and textures to match just about any decorating scheme. And quality can be chosen according to your budget and the traffic that will be expected in the room.

Until recent years, carpeting woven from natural wool was the standard of quality. No other type of fiber could match wool for warmth, wear, and appearance. But high-speed manufacturing techniques have replaced woven carpeting with tufted carpeting. In addition, a variety of modern synthetic fibers have largely replaced wool.

In this chapter, we'll show you how to select carpeting and plan your project *(page 98)*. On page 99, you'll find the tools you'll need. Beginning on page 101, we'll give you step-by-step instructions to install wall-to-wall carpeting on both floors and stairs. You'll also learn how to repair cuts and tears and how to patch carpeting *(page 109)*. Turn to page 100 for tips on how to keep your carpeting looking clean and new.

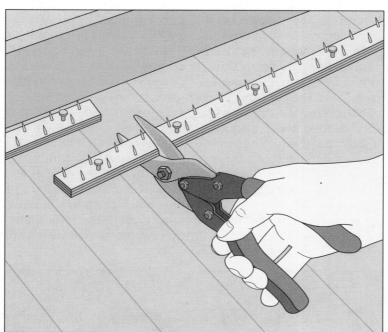

Conventional carpeting is installed using "tackless strips," strips of wood peppered with protruding tacks—so named because they take the place of the individual tacks formerly used in carpet installation.

SELECTING CARPETING

Most carpeting is now made of synthetic fiber, rather than wool. Synthetics have proven to be long-wearing, good-looking, and much less expensive. They're easier to handle than wool and less likely to be damaged during installation. Synthetic fibers also offer the advantage of being nonallergenic.

Synthetics include nylon, polypropylene, polyester, and acrylic. **Nylon** is the dominant fiber on the market. It's a good choice for a high-traffic area because it withstands wear and soiling. **Polypropylene** is used for indoor-outdoor carpet because it is moisture- and mildew-resistant. It's also colorfast, wear-resistant, and easy to clean. **Acrylic** resembles wool in look and feel, except for a slight sheen. It's harder to clean than some other fibers, and tends to shed and become crushed under traffic. **Polyester** offers bright colors and a soft, luxurious feel, but tends to "pill" and crush down. Except for water-soluble stains, it can be hard to clean.

Regardless of the type of fiber used, synthetic carpeting is manufactured either with or without a cushion backing. Carpeting without a cushion backing is commonly referred to as conventional; carpeting with a bonded backing is called attached-cushion-back.

Conventional carpeting is usually installed under tension with stretchers and secured around the perimeter of a room with tackless strips. It can also be installed using double-sided tape or simply laid loose, like an area rug with the edges bound to keep them from fraying. Stretching gives by far the best results. Conventional carpeting usually requires a separate pad, not only to keep it from wearing more quickly, but also to cushion the surface.

Attached-cushion-back carpeting is installed either with adhesive or double-sided tape. No stretchers or tackless strips are required. Although this type of carpeting is relatively easy to install, it is extremely difficult to remove; also, it may tend to bubble and won't wear as well as conventional carpeting. Attached-cushion-back carpeting will not be covered in this book.

All carpeting is made up of tufts; they may be cut or looped *(below)*. Whatever the type of tuft, the pile will tend to fall in one direction. To check the pile's direction, run your hand over it, parallel to the length of the carpet. In one direction, the pile will stand up and look darker; in the other, it will lie flat and look lighter.

ORDERING CARPETING

Careful planning before ordering will help you avoid ending up with carpeting that doesn't fit properly or with the expense of ordering more than you need.

Drawing a floor plan: Prepare a scale drawing of the area to be carpeted, noting exact measurements and marking doorways and other obstacles. These measurements will help you and your dealer select the carpeting that fits your needs most economically. You'll also need the floor plan to cut the carpet later on.

Arranging the layout: Selecting carpeting will be easy if you're fortunate enough to be able to cover your floor with a single piece. But if you'll need seams, here are a few special points to consider in locating them: Because seams are the weak points in wall-to-wall carpeting, you'll want them where they're least visible and away from the room's most heavily trafficked area. Seams also tend to be less visible when they run in the direction of a room's primary outside light source.

Because carpeting generally will look the richest and show its color to best advantage when you look into the pile, try to have the pile pointing in the direction from which it will be viewed most often—toward the doorway, for example. The pile on all the pieces must run in the same direction. Don't overlook the need to match patterns; a special allowance has to be made in figuring the right amount of carpeting in this case.

Estimating cushion: All carpeting feels more comfortable underfoot if it has padding, known as cushion. Carpeting installed with a cushion also wears longer. To come up with the most economical buy, ask your carpet dealer to use your floor plan to figure out exactly how much cushion you'll need. Cushion should not be more than $1/2$ inch thick, and should be firm and dense. Common materials include rubber, vinyl, and urethane.

TUFTED TEXTURES

Loop pile
Withstands traffic well. Level loop gives a flat appearance; multilevel loop surfaces, where loops are of different heights, give a textured appearance.

Cut pile
Offers a variety of textures, commonly with a formal-looking plush surface.

Cut and loop
Some tufts are cut and others are looped to create designs on the surface.

Illustrations courtesy of the Carpet and Rug Institute

BEFORE YOU BEGIN

Some of the common tools you'll need to install your carpeting include a screwdriver, tape measure, and a claw hammer. If you're fastening your carpeting to a concrete slab, you'll need a ball-peen hammer for nailing. The more specialized tools you'll need are shown below; most can be rented from your carpet dealer. Additional tools may be needed to prepare your floor or subfloor; see the instructions on the next page. Basic carpentry skills needed for this preparatory work are not covered in this book.

Make sure you have the necessary safety equipment on hand. You should wear eye protection when you're using a striking tool, especially if you're fastening into concrete. You may want to wear work gloves when handling the tackless strips.

To secure conventional wall-to-wall carpeting laid under tension, you'll need tackless strips made from 1/4-inch lightweight plywood, 1 inch wide and 4 feet long. Each strip is peppered with sharp tacks or pins sticking up at about a 60° angle from the face of the strip. The tacks are 3/16, 7/32, or 1/4 inch long. Ask your dealer which length you need for the carpet you're installing. The strips come equipped with nails already in place for attaching to a wood floor, or with masonry nails for use on concrete floors. You can also glue the strips to surfaces that can't be nailed.

You'll need binder bars to hold the edges of the carpeting in place in doorways. You may also need duct tape for piecing cushion together, hot-melt or latex tape and adhesive for seaming, and carpet tacks for use on stairways.

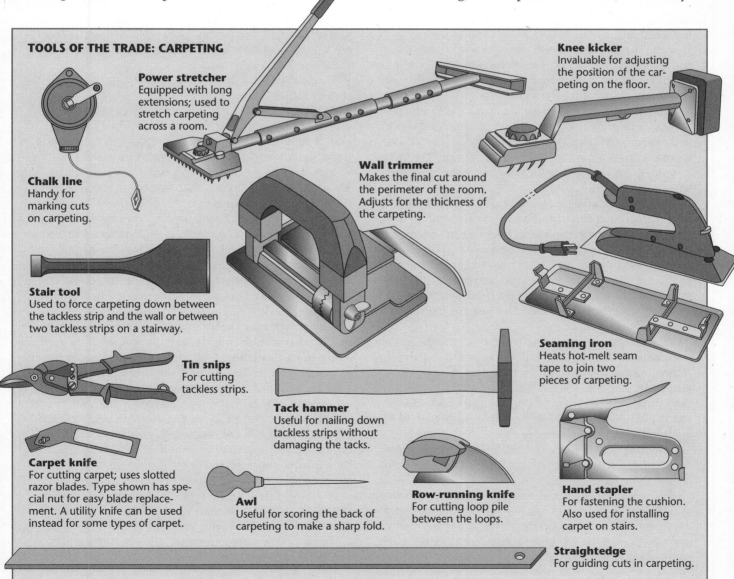

TOOLS OF THE TRADE: CARPETING

Power stretcher
Equipped with long extensions; used to stretch carpeting across a room.

Knee kicker
Invaluable for adjusting the position of the carpeting on the floor.

Chalk line
Handy for marking cuts on carpeting.

Wall trimmer
Makes the final cut around the perimeter of the room. Adjusts for the thickness of the carpeting.

Stair tool
Used to force carpeting down between the tackless strip and the wall or between two tackless strips on a stairway.

Seaming iron
Heats hot-melt seam tape to join two pieces of carpeting.

Tin snips
For cutting tackless strips.

Tack hammer
Useful for nailing down tackless strips without damaging the tacks.

Carpet knife
For cutting carpet; uses slotted razor blades. Type shown has special nut for easy blade replacement. A utility knife can be used instead for some types of carpet.

Awl
Useful for scoring the back of carpeting to make a sharp fold.

Row-running knife
For cutting loop pile between the loops.

Hand stapler
For fastening the cushion. Also used for installing carpet on stairs.

Straightedge
For guiding cuts in carpeting.

PREPARING THE SURFACE

One of the appealing characteristics of conventional carpeting is that little work is necessary to prepare the flooring underneath. Carpeting can be laid over almost any existing floor or subfloor, as long as it's clean and reasonably smooth.

Before installing carpeting, thoroughly examine the floor or stairs and supporting structure from above and below, as described on page 21. Fix any minor problems—protruding nails, loose boards or tiles, minor squeaks, or an occasional hole or gap. Keep in mind that small indentations and irregularities in the flooring will be absorbed by the carpet cushion. Also clean the floor thoroughly. If you're installing carpet over a concrete slab, test it for moisture as described on page 32. Carpet should never be installed over a damp slab. For extra moisture protection on a below-grade slab, use a rubber pad.

Carpet should not be laid over old carpet because it encourages mildew and could present a fire hazard.

CARING FOR YOUR CARPETING

Regardless of the fiber used, proper care of carpeting requires regular cleaning—the best form of preventive maintenance. Microscopic dirt particles on the surface do the damage that wears out carpeting. Grit that's not removed from the surface chews deeper into the fabric as the carpet is walked on.

Vacuuming carpeting once or twice a week, whether the surface looks dirty or not, should be sufficient for removing dirt. Make sure that your vacuum cleaner is in good condition; empty the bag regularly.

Even if you vacuum regularly an overall cleaning is necessary once in a while to get rid of accumulated dirt and restore a carpet to its original brightness. You can hire a professional carpet cleaner, or do it yourself. The three basic methods you can use are hot-water extraction, wet shampoo, and dry cleaning.

Hot-water extraction: Because it doesn't employ mechanical brushing that can distort pile and cause flaring of yarn tufts, hot-water extraction (also called steam cleaning) is generally recommended as the best method for cleaning most types of carpeting. The system forces hot water into the carpet, then vacuums the water and dirt out. The greatest risk of damage with this system is over-wetting; too much water left in the carpet can cause the backing to shrink, mildew, or stain.

Wet shampoo: Many do-it-yourselfers use this system because the cleaning units are widely available for rent. A rotary brush works a detergent foam into the carpeting to loosen dirt particles. The dirt is suspended in the foam while the solution dries. When it's dry, the carpet is vacuumed. Use the minimum of detergent possible, so that less residue is left in the carpet. Any residue causes the carpeting to get soiled faster, and leaves a dull finish. With this method, there is some chance of over-wetting, and the brushes may distort some types of pile.

Dry cleaning: Powdered compounds are worked down into the pile of the carpet, usually with a dry carpet cleaning machine. The loosened dirt and powder are then vacuumed. As with the wet shampooing method, there is some risk of distorting the pile.

Removing stains: Modern synthetic-fiber carpeting is treated with special finishes to resist staining and soiling. These finishes hold the spilled material on the surface of the carpet, giving you time to mop up or remove it before it penetrates deeper into the fibers. Of course, the longer you leave a spill, the harder it will be to remove it. There are a number of cleaning methods you can use to remove stubborn stains. They include:
• Dry cleaning fluid (available at grocery and hardware stores); cleans butter or grease stains.
• Nail polish remover (do not use if it contains acetone); useful for ink, paint, and shoe polish stains.
• Mild detergent solution (2 cups water and 2 cups mild liquid detergent); removes food and drink stains.
• Lukewarm water; use for most types of stains.
• Vinegar solution (1 cup white vinegar to 2 cups water); recommended for rust stains.
• Ammonia solution (2 tablespoons household ammonia to 1 cup water); use for more stubborn food and drink stains.
• Commercial spot removal kits (available from carpet dealers and professional carpet cleaners); usually a last resort.

Contact the carpet manufacturer or the Carpet and Rug Institute in Dalton, Georgia, for more detailed stain removal guidelines.

Test any cleaning solution on an inconspicuous area first, to ensure it will not damage the carpet or its dyes. Then, use a clean, white, absorbent cloth or paper towel to gently work in the cleaning agent; start from the edges of the spot and work toward the center. Blot the area thoroughly until the stain no longer transfers to the cleaning cloth. Use clean water to remove any cleaning residue and absorb the remaining moisture with several layers of paper towel. Wait until the carpeting is dry, then brush or vacuum it. For stains that resist all your cleaning efforts, you'll have to call on the services of a professional carpet cleaner. NOTE: Some spills contain chemicals that may discolor or even damage the carpet fibers or dyes. If you're not sure what caused the spot and don't know how to remove it, it's best to contact a professional cleaner.

INSTALLING CONVENTIONAL CARPETING

Once you're satisfied that the surface of the floor or subfloor is dry, clean, and in good shape, you can begin the actual work of installing the carpet. First, remove any shoe molding *(page 36)* from around the baseboards and any floor grates or other obstructions. Number the pieces of molding with chalk or pencil if you're planning to replace them. If you're installing very thick carpeting, you may need to remove and trim the doors, even if they open out of the room; carpeting will extend under the door. Finally, make sure that the pile on all pieces of carpeting lies in the same direction.

Preparing to lay carpet

TOOLKIT
- Tin snips
- Tape measure
- Tack hammer for fastening into wood
- Ball-peen hammer for fastening into concrete
- Hand stapler
- Utility knife

1 ▶ **Installing tackless strips**
Begin by placing lengths of tackless strips in a continuous line around the perimeter of the room; make sure the tacks face the wall *(right, inset)*. You can either cut the strips to fit now or cut and nail as you work. Use tin snips to make the cuts in the soft wood *(right, above)*.

When you're ready to nail the strips in place, position the first one so its distance from the wall is slightly less than the thickness of the carpet but not more than $3/8$". Once you determine the distance, make a spacer from a scrap of wood or cardboard *(right, below)* so you can position the remaining strips uniformly.

When nailing the strips to the floor, use a tack hammer or a hammer with a small head to avoid damaging the tacks with any misdirected blows. If you're fastening to resilient flooring with wood beneath it, nail right through the old flooring into the wood. For driving masonry nails into a concrete subfloor, you'll need a ball-peen hammer. Be sure to wear safety goggles while driving nails into concrete.

To fasten tackless strips to ceramic tile or any other floor covering that can't be nailed, use contact cement. To get a good bond, follow the directions of the manufacturer and observe all recommended safety precautions.

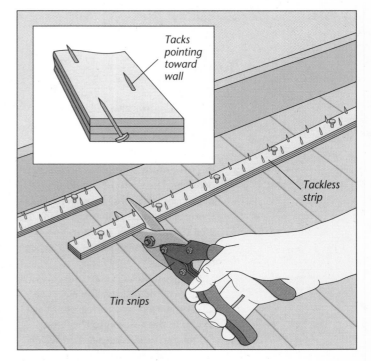

Tacks pointing toward wall

Tackless strip

Tin snips

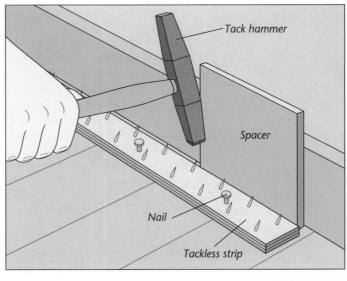

Tack hammer

Spacer

Nail

Tackless strip

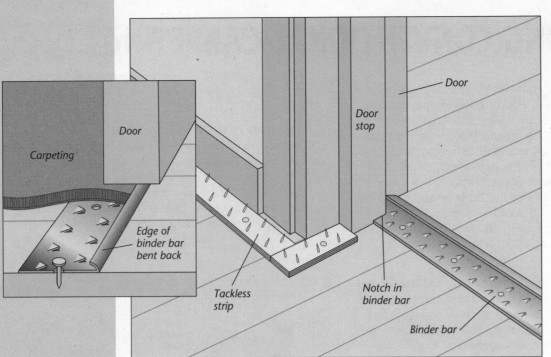

Carpeting

Door

Edge of binder bar bent back

Door

Door stop

Tackless strip

Notch in binder bar

Door

Binder bar

◀ 2 **Installing binder bars**

Attach binder bars in doorways or any other places where the carpeting ends, except against walls. Position each bar so the bent edge will be centered under the door when it's closed. After the carpeting is laid, the raised edge of the bar will be bent down over the edge of the carpet. If the door opens away from the carpeted area, you may have to notch the bar with tin snips to accommodate the door's stop molding, as shown at left.

3 ▶ **Laying the cushion**

When the tackless strips and binder bars are in place, unroll the cushion and cut it so that it covers the maximum amount of floor area possible. The cushion should overlap the tackless strips just enough to ensure there's adequate cushion to fill up the area to the strips; you'll trim it later. If the cushion has a smooth side, place this side up.

If your subfloor is concrete, masonry, or ceramic tile, use the adhesive recommended by your carpet dealer to secure the cushion to the floor. Otherwise, work along the edges of the tackless strips and fasten the cushion with staples placed every 6" to 8" *(right, top)*.

Cut additional pieces of cushion as necessary to cover the entire floor, taping them together with duct tape and securing the edges with staples. Also staple any areas where you think the cushion may slip as you lay the carpet.

When all the cushion is securely in place, take a sharp utility knife and trim off the overlap along the inside edge of the tackless strip *(right, bottom)*.

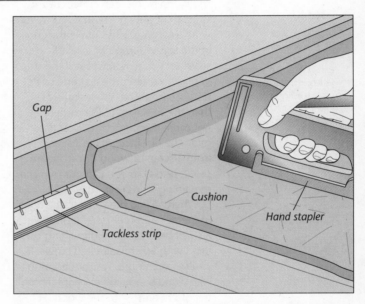

Gap

Cushion

Hand stapler

Tackless strip

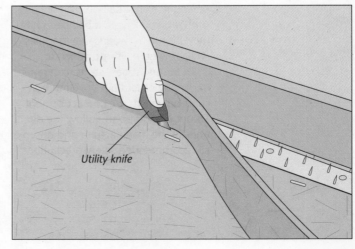

Utility knife

TOOLKIT
For both types:
• Straightedge
• Seaming iron
• Chalk line
• Carpet knife
For loop pile:
• Screwdriver
• Row-running knife

1 Making the rough cuts

Using the scale drawing you prepared to determine your original carpeting order, roll out the carpet and measure for the cuts you have to make. (If you don't have a large enough area indoors, use a clean, flat, and dry area outdoors.)

Measure carefully before making any cuts. Allow at least a 3" overlap around the edges of the room and a couple of extra inches for trimming any seams. Make sure that patterns will match and that the pile runs in the same direction.

For these rough cuts, you can cut both loop and cut pile carpet from the back. First, notch the face of the carpet on both edges to indicate the location of the cut. Then fold the carpet back and snap a chalk line between the notches. Using a straightedge and a carpet knife, cut along the chalk line, being careful to cut only through the carpet's backing.

Cutouts for registers should always be made after the carpet is stretched.

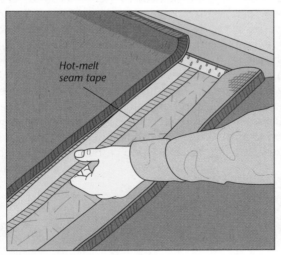

1" overlap

Row-running knife

2 Cutting a seam

Never use the end of a roll of carpeting for a seam—always make a straight cut on both pieces.

For loop pile, make the cuts from the front. Cut the first piece by clearing a path in the pile with a screwdriver and then using a row-running knife to cut between the rows of loops. Next, place the first piece so that it overlaps the second piece by 1". Place the pieces so the top piece has the nap leaning over the seam. Check to see that the top piece is straight. Then, take a row-running knife and cut the bottom piece of carpet using the edge of the top piece of carpet as a guide *(left)*. Cut as close to the wall as you can with the row-running knife, then complete the cut with a carpet knife.

For cut pile, measure both pieces carefully and cut them from the back as described above.

Hot-melt seam tape

Seaming iron

Tape

3 Sealing the seam

Take a length of hot-melt seam tape and slip it, adhesive side up, halfway under one carpet edge *(above, left)*.

Holding back one edge of the carpet, slip the pre-heated seaming iron under the other carpet edge and hold it on the tape, leaving it in place for about 30 seconds. Then draw the iron slowly along the tape as you press the carpet edges down into the heated

adhesive *(above, right)*. Check to see that the two carpet pieces are butting. If they're not, press them together and place a clean, heavy object (not metal) on the seam to keep the pieces from pulling apart. Continue working across the floor to where the carpet edges rest against the wall. Pull the carpet away from the wall to complete the seam. Allow the seam to set until cool before stretching the carpet.

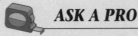

ASK A PRO

HOW DO I USE LATEX TAPE?

You can use latex tape instead of hot-melt tape. However, the seam will be more visible and won't hold as well under heavy traffic.

Cut strips of tape to fit the entire seam, coat them thinly with adhesive, and center them under the carpet edges. Squeeze a thin bead of adhesive along the edges of the backing and not on the pile. Then, press the carpet firmly onto the tape, blending the pile with your fingertips. Let the adhesive dry thoroughly.

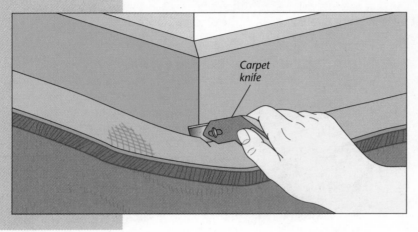

Carpet knife

4 Preparing the carpet for stretching

Once any seams are thoroughly set, you're ready to begin stretching the carpet. Walk around the perimeter of the room and use your feet to shift the carpet so it lies smoothly. Then, with a carpet knife, trim away any excessive overlap; the carpet should overlap the tackless strips by 1" or 2".

Make small relief cuts with a carpet knife at each outside corner of the room *(left)* so the carpeting will rest flat on the floor. Cut the carpet from the back as shown. Cut around other obstacles in the same way, making as many cuts as necessary to allow the carpet to lie flat.

STRETCHING TOOLS: TAKING UP THE SLACK

A knee kicker is used to slip the carpet up and over the tackless strips. A power stretcher pulls the carpet across the room so it can be secured on the opposite strips.

The knee kicker *(below, left)* has a head with adjustable teeth. Small hooks catch the nap as longer teeth reach down to grip the backing. The opposite end is padded so you can use your knee to bump the tool forward; the handle telescopes. To use the knee kicker, place its head about an inch from the tackless strip. When you bump the pad with your knee, the head will move forward and catch the carpet backing on the pins of the tackless strip, holding the carpeting in place.

The head of the power stretcher *(below, right)* is similar, but larger; adjustable teeth protrude from the head to grip the carpeting. A series of handle extensions allows the power stretcher to be braced against an opposite wall; when the tool is in place, you lower a lever that moves the head forward with steady pressure.

Experiment to see how much bite is necessary to get a good grip on the carpet. Stretch the carpet with the minimum force necessary to pull the carpet taut; too much pressure can tear it. Protect the baseboards by placing a block of wood cushioned with a piece of carpet between the back of the stretcher and the baseboard.

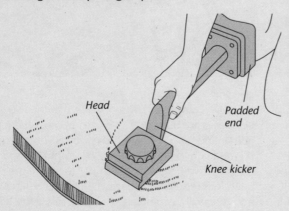

Head

Padded end

Knee kicker

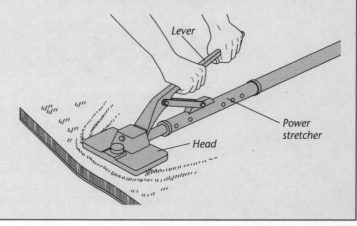

Lever

Power stretcher

Head

Stretching and securing the carpet

TOOLKIT
- Knee kicker
- Power stretcher
- Wall trimmer
- Carpet knife
- Stair tool
- Hammer

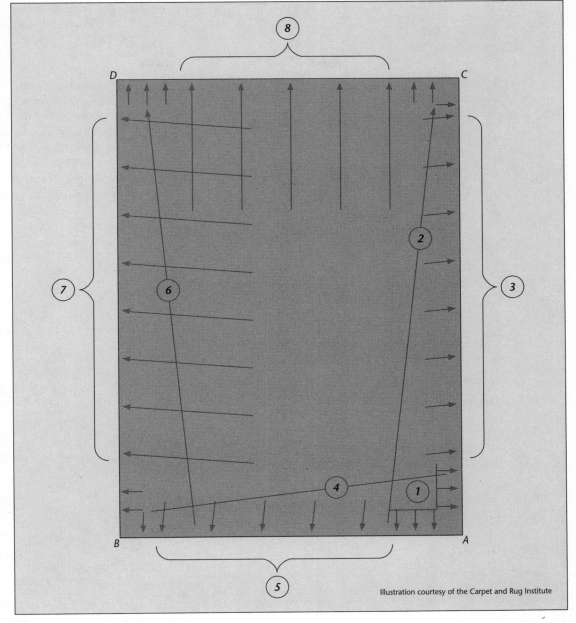

Illustration courtesy of the Carpet and Rug Institute

1 Stretching the carpet

Refer to the illustration above for the eight steps to stretch your carpet. NOTE: The short arrows indicate the use of a knee kicker and the medium and longer arrows indicate the use of a power stretcher.

(1) Hook the carpet along corner A (above), using the knee kicker (about 2' in each direction).

(2) Power-stretch from wall AB and hook at corner C.

(3) Stretch and hook the carpet with the knee kicker along the wall from A to C. Work at a slight angle so that the carpet is stretched toward corner C.

(4) Power-stretch from wall AC and hook at corner B.

(5) Stretch and hook with the knee kicker along the wall from A to B, working at a slight angle as shown.

(6) Power-stretch from wall AB and then hook the carpet temporarily at corner D until Step 7 has been completed.

(7) Power-stretch from wall AC and hook along the wall from B to D.

(8) Power-stretch from wall AB and hook along the wall from C to D.

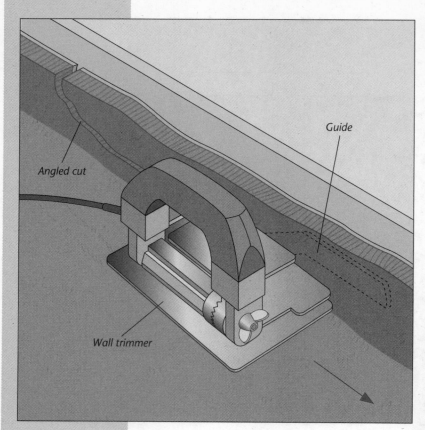

Angled cut

Guide

Wall trimmer

2 ▶ Trimming the carpet at the walls

If possible, use a wall trimmer to make the final cut around the perimeter of a room; otherwise, use a carpet knife for all cuts, cutting from the back.

Before using the trimmer, first slice downward into the carpet at an angle with a knife. Then, adjust the trimmer for the thickness of the carpet. Leaving just enough carpet to be folded over the tackless strip and down into the gap between the strip and the wall (about ¼"), guide the trimmer along the wall *(left)*. In corners and around obstacles, use a carpet knife to make the cuts from the back.

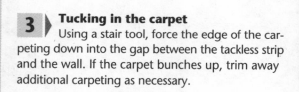

3 ▶ Tucking in the carpet

Using a stair tool, force the edge of the carpeting down into the gap between the tackless strip and the wall. If the carpet bunches up, trim away additional carpeting as necessary.

Stair tool

Carpet

Tackless strip

Cushion

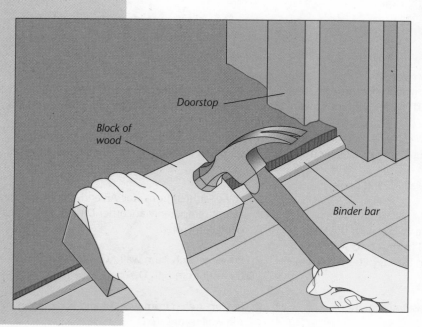

Doorstop

Block of wood

Binder bar

4 ▶ Finishing the job

Where you've used binder bars, trim the carpet to fit; then take a block of wood and a hammer and rap the top edge of the binder bar to bend it tightly over the edge of the carpet.

To cut holes for registers, find the hole and insert your carpet knife into the center. Cut toward the edge and then follow around the edge of the hole.

Although most wall-to-wall carpeting is installed without shoe molding, you may decide to replace the shoe molding if you feel it adds to the appearance.

INSTALLING CARPETING ON STAIRS

Many homeowners with wall-to-wall carpeting prefer to extend carpeting up stairways for aesthetic reasons. But there are practical reasons for carpeting stairs as well. Carpeted stairs are quieter and safer, reducing the likelihood of falls and providing a cushioned surface should such a mishap occur.

Enclosed stairways can be carpeted wall to wall. Stairways with one open side can be carpeted from the wall to the balusters on the open side; the carpet can also be cut around the balusters, but then it will have to be folded around the edge of the stairs. Alternatively, a couple of inches of wood can be left exposed along the edge of the stairs. This method is shown below. If you're carpeting up to a baluster or leaving part of the stair uncarpeted, you'll need to fold the exposed edge under 1¼ inches. If you're carpeting up to a wall you'll need to install tackless strips along the wall; in this case the edges won't be folded under.

Placing and attaching the carpet

TOOLKIT
- Tape measure
- Tin snips
- Tack hammer
- Utility knife
- Hand stapler
- Carpet knife for cut pile or row-running knife for loop pile
- Chalk line
- Straightedge
- Awl
- Stair tool
- Knee kicker
- Hand stapler

1 ▶ Measuring the stairway
To determine how much carpeting you'll need, measure the distance from the back of one tread to the bottom of the riser below. Add another inch to allow for padding under the carpet. Multiply that figure by the number of steps to get the total length required. Measure the width of the part of the stair you're carpeting.

Remember to add a total of 2½" if you'll be folding the edges under. Also, measure for any landing at the top of the stairs.

Take the measurements with you to your carpeting supplier, who can recommend how much you need to buy and how to make the most economical cuts.

2 ▶ Installing tackless strips
Install tackless strips at the back of each tread and the bottom of each riser. Cut each piece 2½" shorter than the width you're covering to allow for folded-under borders. Use a scrap of wood ⅝" thick as a spacer. Working down the stairway, rest the first piece of tackless strip on the spacer and nail it ⅝" up from the tread; be sure that it's centered on the face of the riser and that the pins point down.

Nail a second strip (also centered) near the back of the top tread ⅝" out from the riser (pins should point toward the riser). Continue adding strips down the stairs. If your stairway has a landing, carpet it in the same way wall-to-wall carpeting is installed in a room; leave enough of an overhang on the down side of the landing so the carpeting can be wrapped down to cover the riser of the step below.

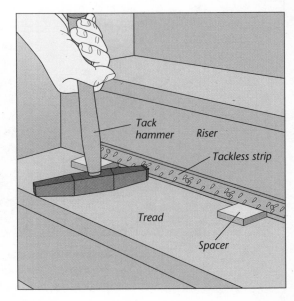

Tack hammer   Riser

Tackless strip

Tread

Spacer

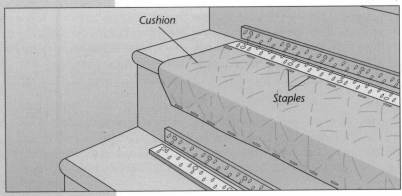

Cushion

Staples

3 ◀ Installing the cushion
Individual pieces of cushion must be cut and stapled to each step. Use a utility knife to cut each piece of cushion to the same width as the tackless strips. Butt the cushion against the strip at the back of the tread, then fold it forward over the edge of the tread, cutting it about 3" above the tackless strip on the riser below. Staple the cushion to the stairs along the top and bottom edges as shown.

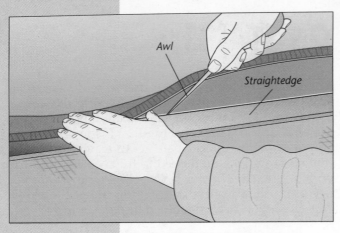

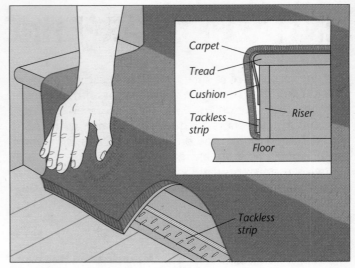

4 Cutting and folding the edges

Cut the carpeting into a strip or strips of the proper width. (See page 103 for instructions on cutting carpet.) Place each strip face down on a flat working area. Snap a chalk line along each edge of the carpeting 1¼" from the edge. Use an awl and a straightedge to score the back of the carpet *(above)*. As it's scored, fold the edges over. When the edges have been folded the full length of each strip, place the carpet into position.

5 Beginning at the bottom step

With the pile running down the stairs, center the carpet on the stairs with the lower edge overlapping the floor about one carpet thickness and the folded-under edges falling outside the ends of the tackless strips. With your hand, ease the carpet down over the pins of the tackless strip on the bottom of the riser *(above)*. Press the carpet onto the pins. Then drive a carpet tack through the folded-under edge on each side and into the riser.

Use a stair tool to tuck the bottom edge of the carpet into the gap between the tackless strip and the floor.

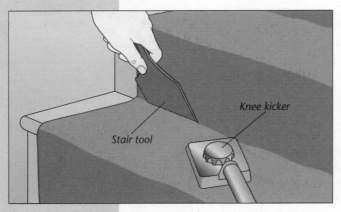

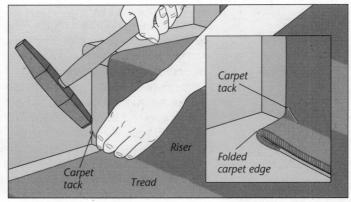

6 Fastening the carpet

When the lower edge of the carpet is secure, place the head of the knee kicker in the center of the tread and bump the carpet taut. At the same time, use a stair tool to force the carpeting down into the space between the two tackless strips, starting in the middle of the step *(above, left)*. Once the carpet is secured in the center, angle the knee kicker toward each side of the step and move across the step, securing the carpeting. Then drive a carpet tack through the folded edge on each side of the step into the corner where the tread and riser meet *(above, right)*. Stretch and secure the carpeting one step at a time. If you want to join two strips of carpeting, do so at the tread/riser joint.

7 Finishing the top

If the carpeting ends at the top of the stairs, carpet the last riser, ending under the nosing: Tuck the carpet into the tackless strips at the bottom of the last riser only partway. Then cut the carpet along the top of the riser, leaving about ¼" of extra material. Staple the carpet under the nosing and then take up the excess by tucking the carpet the rest of the way into the tackless strips below.

REPAIRING CARPETING

Although usually made of durable synthetics, carpeting is one of the more fragile floor coverings; it's vulnerable to snags, enduring stains, cigarette burns, and excessive wear in heavy-traffic areas.

You can make many carpet repairs yourself, but before beginning any repair job, it's wise to check for underlying causes of carpet damage below the floor surface. A thin strip of worn carpeting might indicate loose or warped subflooring. Loose, protruding nails can cause an obvious pattern of worn dots in the carpeting above. For instructions on how to check the basic floor structure and for suggested remedies to structural problems, see page 21.

Most major carpet damage—burns, stains, or ragged tears—can be repaired by removing the damaged section and replacing it with a new piece. The trick is to find a replacement piece that will match the surrounding carpet. The ideal replacement is leftover material from the original installation. However, if your old carpet is faded or worn, the replacement may no longer match exactly. If you don't have leftovers, check with carpet suppliers to see if the particular type and color of carpet you have is still available. Try to take along a scrap of the original carpet; otherwise, it may take time to find the right match. If you can't find anything acceptable, consider taking a section of original carpet from a closet or another out-of-the-way area; replace it with a less suitable piece. In this section, we'll give you instructions for patching and stitching minor tears and cuts in conventional carpeting.

TOOLS AND SUPPLIES

To stitch up minor cuts or tears in carpeting, you'll need a knee kicker, an awl, a heavy curved needle, carpet thread that matches the color of your carpet, and latex carpet adhesive.

For patching carpeting, you'll need a knee kicker, a carpet knife or utility knife, a hammer, carpet tacks, and carpet seam tape. Carpet seam tape is available either precoated or uncoated—ask your flooring dealer which is the best for your situation; if you choose uncoated, you'll also need latex seam adhesive.

A knee kicker (page 104) can be rented from a carpet dealer. The other supplies can usually be purchased from the same source. For illustrations of the specialized tools that you'll need, turn to page 99.

Repairing minor cuts or tears

TOOLKIT
• Knee kicker
• Awl
• Carpet knife
• Heavy curved needle

Repairing a tear

You can use carpet seam tape to repair minor cuts or tears, or stitch up the tear with carpet thread reinforced with latex seam adhesive. With either method, use the knee kicker to reduce tension in the carpet in the corner of the room closest to the tear. Use an awl to free the carpet from the tackless strips along the edges nearest the tear (right). Pull back the carpet to expose the underside of the tear. Before stitching or seaming, trim the edges. Use a heavy needle and carpet thread to sew the tear together; make the stitches approximately 1" long and 1/4" inch apart. Run the needle up through the carpet backing and into the pile (inset), and cover the bottom of the stitched area with latex carpet adhesive. Place a piece of absorbent tissue (such as toilet paper) over the adhesive, and reattach the carpeting to the tackless strips with the knee kicker. For large tears, or for tears far from a corner, replace the torn carpet with a patch, as described on page 110.

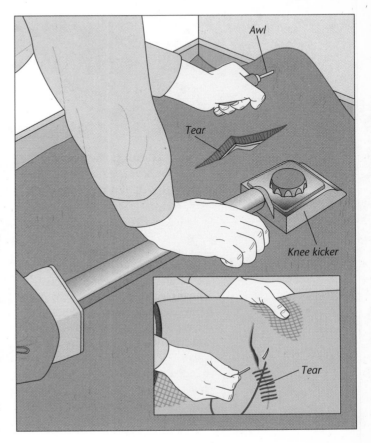

Awl

Tear

Knee kicker

Tear

Patching carpet

TOOLKIT
- Knee kicker
- Tack hammer
- Carpet knife for cut pile or row-running knife for loop pile
- Prybar

1 ▶ Setting up

Most conventional carpeting is installed under tension, which you'll have to release. Set a knee kicker about 1" from the area, and nudge it forward until any tension in the area has been released. To keep the area you're working on loose, tack down the carpet around it. Use strips of old carpet 2" wide and about 2" longer than the cuts you plan to make around the damaged area. Place the first strip upside down in front of the knee kicker, and tack it in place. Using the knee kicker in the same way on each side, tack the remaining carpet strips in a square around the damaged area (right).

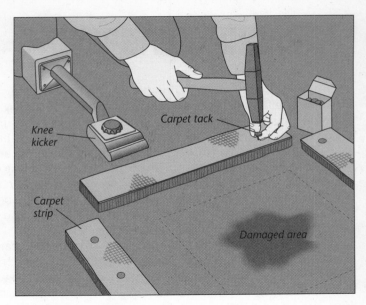

Knee kicker

Carpet tack

Carpet strip

Damaged area

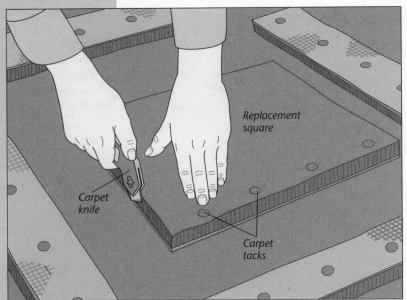

Replacement square

Carpet knife

Carpet tacks

2 ◀ Cutting the patch

From a piece of replacement carpet, cut out a square slightly larger than the damaged section, matching any pattern and direction of pile. Place it over the damaged area, right side up. (To cut carpet, see page 103.) If your carpet is laid over wood, drive nails or long carpet tacks along one edge of the new piece, through the old carpet, and into the floor to keep both the old and new carpet in place; you'll need to remove them later, so don't drive them too hard. If your carpet is not laid over wood, cut one side of the old carpet first and pry it up with your fingers. Use double-sided tape to secure the damaged part of the old carpet to the new piece. Using the new piece as a guide, cut out the damaged area, taking care to cut between the rows of pile (left). Don't cut into any cushion; if you do, mend it with duct tape. Pry out the nails or tacks from the replacement square.

3 ▶ Fastening the patch

Remove the damaged square and cut lengths of carpet seam tape to fit along the edges of the original carpet. If you're using uncoated tape, cover half of each strip with latex seam adhesive. Place the coated edge of each strip, sticky side up, under the edge of the original carpet (right). Precoated tape is placed in the same way. Apply more seam adhesive, if necessary, to the exposed portion of the tape, and press the new piece of carpeting into place. Check the drying time recommended by the adhesive manufacturer; allow enough time for the adhesive to dry before removing the carpet tacks from the carpet strips.

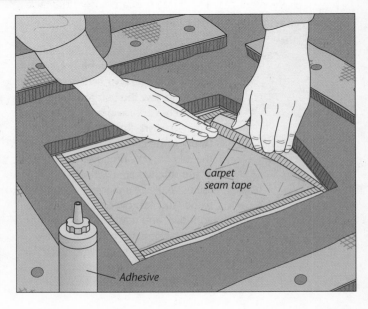

Carpet seam tape

Adhesive

FLOORING GLOSSARY

Asphalt felt
Felt saturated with asphalt; used as a moisture barrier.

Batten
A wood strip, usually 1x2 or 1x3, that gives a rigid guide to set tile against.

Bedding
Gently tapping a tile or parquet square to set it firmly in the adhesive and level with adjoining tiles.

Blind-nailing
Nailing at an angle so the nailhead is hidden. Tongue-and-groove strips and planks are blind-nailed through the tongue.

Bond
The adherence of one material to another, such as between tile and adhesive. Also, a pattern of joint alignment for brick or tile, as in jack-on-jack or running bond.

Bridging
Wood or metal in X-shape or solid wood block inserted between joists to prevent them from twisting.

Casing
Decorative trim around a door or window.

Cement, portland
A manufactured product, as opposed to natural cement. A basic ingredient in mortar.

Chalk line
A reference line marked on a surface by snapping a string coated with chalk.

Cleat
A piece of lumber used as a support or to keep something in place.

Conventional carpeting
Carpeting that comes without an attached cushion. Installed by stretching onto tackless strips.

Counterboring
Drilling a hole for a screw at a depth that leaves the head of the screw below the surface of the wood. Allows for the hole to be plugged, concealing the screw head.

Countersinking
Drilling a hole in wood to allow a flathead screw to sit flush with the surface. Can be drilled inside a counterbored hole.

Cove base
A flexible vinyl or rubber trim strip often 4" high, installed along the base of a wall. May be used with resilient flooring instead of baseboards and shoe molding.

Curing
The process by which mortar or adhesive sets. Mortar needs to be kept damp ("damp cure"), while most adhesives cure as they dry ("dry cure").

Dry run
Laying tile out on the floor before applying adhesive, to determine the best layout and minimize the number of cut tiles.

Face-nailing
Fastening a board by nailing straight down through it into the material below.

Flagstone
Rough-hewn stone split in layers. May be one of a number of types of stone, commonly sandstone.

Footing
A concrete base for a post, made wider than the post to spread the weight.

Girder
A heavy, horizontal framing member supporting floor joists, resting on either columns or in pockets in the foundation walls.

Glaze
Hard, glassy coating fused to the top surface of a ceramic tile by firing at a high temperature.

Grade
Ground level. Floors are either on grade, below grade, or above grade.

Grout
A material used to fill the joints between ceramic tiles to keep out dirt and liquids.

Jamb
A board that forms the top or side of a frame surrounding a door or window.

Joist
A horizontal framing member placed on edge. Floor joists support the subfloor.

Laminate flooring
Wood flooring, either parquet or plank, made up of more than one layer of wood veneer.

Mastic
A glue-like adhesive.

Miter cut
An angled cut at other than 90°, typically 45°.

Moisture barrier
Material such as polyethylene film or asphalt felt laid over a subfloor to prevent moisture from penetrating up into the finished floor.

Mortar
Used in a thick bed to bond masonry flooring to the subfloor. Composed of portland cement, sand, and lime. Thin-set adhesives may also be referred to as mortar.

Open time
Time during which an adhesive retains its ability to stick to a tile and bond it to the backing.

Parquet
Wood flooring made in squares or rectangular blocks. Also referred to as wood tile or block.

Penetrating finish
A wood finish that penetrates the pores of the wood, leaving a matte surface.

Pier
A concrete base for a post.

Pile
The tufts of yarn, either cut or looped, that form the surface of carpeting.

Plank flooring
Wood flooring boards in varying widths and random lengths; wider than strips. Usually tongue-and-groove.

Polyethylene film
A thin sheet of plastic used as a moisture barrier.

Post
Wooden or steel column resting on a pier and supporting the girder.

Reducer strip
A strip of wood with a curved edge used to form the transition between wood flooring and adjacent flooring.

Resilient flooring
Flooring made of vinyl, vinyl composition, rubber, or cork.

Ring-shank nails
Nails with rings that provide extra grip. Used for nailing down plywood underlayment.

Sealer
A product used to protect wood, ceramic, and masonry floors from water penetration and stains.

Shim
A thin piece of wood or other material that is driven behind a surface to make it flat or level.

Shoe molding
A strip of wood molding, typically quarter-round, that covers the gap between the baseboard or wall and the flooring.

Silicone rubber caulk
A product used to fill joints, such as between a tile floor and the walls. Remains flexible and waterproof, and resists mildew and extremes of cold and heat.

Sleeper
A length of lumber laid over a concrete subfloor to provide a nailing surface for flooring materials.

Splits
Bricks that are half as thick as standard bricks. Can be used instead of standard bricks for a lighter floor.

Strip flooring
Narrow tongue-and-groove boards of fixed width but random length.

Subfloor
The first layer of the floor structure—either plywood or boards fastened to the joists, or a concrete slab.

Surface finish
A wood finish that forms a clear coating on the surface of the wood.

Tack cloth
Used to remove dust or debris from wood prior to finishing. Store-bought or made with lint-free cloth. Dry or moistened with mineral spirits, depending on finish.

Tackless strips
Strips of plywood with projecting tacks that are installed around the perimeter of a room to hold conventional carpeting in place; used instead of individual tacks.

Thin-set
A cement-base adhesive. Also referred to as thin-set mortar.

Tongue-and-groove flooring
Wood flooring materials having a tongue along one edge that fits into a groove on the adjacent piece.

Tooling
Compacting a mortar joint using a convex jointer or other convex object.

Underlayment
Material laid over the subfloor to smooth out irregularities and increase rigidity. Usually plywood, or, in some cases, cement backerboard.

Working lines
Lines laid out on the subfloor or underlayment to keep the first course of tile or parquet straight, and to adjust for any irregularities in the walls.

INDEX